Mechanik, Werkstoffe und Konstruktion im Bauwesen

Band 56

Reihe herausgegeben von
Ulrich Knaack, Darmstadt, Deutschland
Jens Schneider, Darmstadt, Deutschland
Johann-Dietrich Wörner, Darmstadt, Deutschland
Stefan Kolling, Gießen, Deutschland

Institutsreihe zu Fortschritten bei Mechanik, Werkstoffen, Konstruktionen, Gebäudehüllen und Tragwerken. Das Institut für Statik und Konstruktion der TU Darmstadt sowie das Institut für Mechanik und Materialforschung der TH Mittelhessen in Gießen bündeln die Forschungs- und Lehraktivitäten in den Bereichen Mechanik, Werkstoffe im Bauwesen, Statik und Dynamik, Glasbau und Fassadentechnik, um einheitliche Grundlagen für werkstoffgerechtes Entwerfen und Konstruieren zu erreichen. Die Institute sind national und international sehr gut vernetzt und kooperieren bei grundlegenden theoretischen Arbeiten und angewandten Forschungsprojekten mit Partnern aus Wissenschaft, Industrie und Verwaltung. Die Forschungsaktivitäten finden sich im gesamten Ingenieurbereich wieder. Sie umfassen die Modellierung von Tragstrukturen zur Erfassung des statischen und dynamischen Verhaltens, die mechanische Modellierung und Computersimulation des Deformations-, Schädigungs- und Versagensverhaltens von Werkstoffen, Bauteilen und Tragstrukturen, die Entwicklung neuer Materialien, Produktionsverfahren und Gebäudetechnologien sowie deren Anwendung im Bauwesen unter Berücksichtigung sicherheitstheoretischer Überlegungen und der Energieeffizienz, konstruktive Aspekte des Umweltschutzes sowie numerische Simulationen von komplexen Stoßvorgängen und Kontaktproblemen in Statik und Dynamik.

Weitere Bände in der Reihe http://www.springer.com/series/13824

Alamir Mohsen

Design to Manufacture of Complex Building Envelopes

Single Layer Envelopes: Mullion-Transom Systems + 3D printed Metal Nodes

Springer Vieweg

Alamir Mohsen
Lithium Architects GmbH
Frankfurt am Main, Germany

ISSN 2512-3238 ISSN 2512-3246 (electronic)
Mechanik, Werkstoffe und Konstruktion im Bauwesen
ISBN 978-3-658-30203-0 ISBN 978-3-658-30204-7 (eBook)
https://doi.org/10.1007/978-3-658-30204-7

This Springer Vieweg imprint is published by the registered company Springer Fachmedien Wiesbaden GmbH part of Springer Nature.
The registered company address is: Abraham-Lincoln-Str. 46, 65189 Wiesbaden, Germany

Acknowledgment

To my family, who have provided me thorough moral and emotional support in my life.

Thank you to my supervisors, Prof. Dr.-Ing Ulrich Knaack and Prof. Dr.-Ing. Habil. Stefan Kolling, for providing guidance and feedback throughout this dissertation.

I would like to thank the following people for helping with this dissertation: Michael Drass, Holger Strauß, Marcel Hörbert, Leonie Hartmann, Lorenz Riedel, Paul Küster, Remmal Rizeq, Wandong Chen, Eren Tatar, and Benjamin Kryts.

I am also grateful to Silke Wallner, university staff.

I would like to thank my editors, Michele Campbell and Tamani Hamburg for their support on the last stage of this dissertation.

Abstract

A new method for the design and engineering of complex façades and their require-ments needs to be explored and applied. This method utilizes the file-to-manufactory concept, which explores the opportunities of parametric design approaches and additive manufac_ turing.

The two are growing trends and offer new possibilities. These parametric design ap_ proaches change how planners/designers perceive building details. Therefore, a new de_ sign and engineering concept is needed to envelop the complexity of architectural geom_ etries. This complexity is increasing, due to the rapid developments in all related aspects, such as façade systems, modeling software, and digital manufacturing techniques.

Because of the increase in the complexity of architectural geometries, traditional proj_ ect management is inadequate. The file-to-manufactory concept is a contemporary way to enhance architectural possibilities and maximize efficiency. This concept needs the exploration of specific BIM, CAD, and additive manufacturing techniques. This disser_ tation explores the needed steps and presents them as a design and engineering concept that suits most of the elaborate façades and envelopes.

The design and engineering concept is divided into digital conception and manufac_ turing conception.

In the digital conception, CAD software is used to define the complex geometry of the project parametrically and to generate all the digital information needed to control the manufacturing process. Through the digital conception, the primary rules of the complex façade structure are defined into categories such as connection type and joining system.

Additive manufacturing or 3D printing is one of the computer-aided production tech_ niques utilized for the manufacturing conception. Therefore, this process is investigated and tested to check its suitability for the façade industry.

Keywords: *complex façades and envelopes, parametric design, additive manufacturing (3D printing), parametric planning concept, parametric node, N-AM, Li3-Method.*

Kurzfassung

Eine neue Methode für die Gestaltung komplexer Fassaden und deren technische Usetzung muss erforscht und nutzbar gemacht werden. Diese neue Methode nutzt den File-to-Manufactory-Ansatz, welcher die Möglichkeiten der parametrischen Planung und der additiven Fertigung ausschöpft.

Beides sind wachsende Trends, die neue Möglichkeiten bieten. Die parametrischen Ansätze verändern, wie die Entwerfer und Planer Gebäudestrukturen und –details wahrnehmen. Daher ist auch ein neues Entwurfs- und Realisierungskonzept erforderlich, um der zunehmenden Komplexität der Architekturgeometrien gerecht werden zu können. Diese Komplexität nimmt im gleichen Maße ebenfalls zu, da sich alle damit verbundenen Aspekte, wie Fassadensysteme, Modellierungssoftware und digitale Fertigungstechniken zeitgleich rasant weiterentwickeln.

Mit der wachsenden Komplexität der architektonischen Geometrien wird der traditionelle Entwurfs- und Planungsprozeß zunehmend unzureichender. Das File-to-Manufactory-Konzept ist hierbei ein zeitgemäßer Weg, um die Möglichkeiten der heutigen Architektur zu erweitern und Ihre Effizienz zu steigern, sowohl im Planungsprozeß, als auch in der Herstellung. Das Konzept erfordert hierbei die Erforschung spezifischer Planungstechniken (z.B. BIM), Planungswerkzeugen (z.B. neueste CAD-Tools und Anwendungen) und neue, digitale Fertigungstechniken (z.B. additive Verfahren, 3D-Drucken). Diese Dissertation untersucht die notwendigen Schritte für die Methode und präsentiert die Ergebnisse als ausformuliertes Entwurfs- und Ingenieurkonzept, das den meisten komplexen Fassaden und Gebäudehüllen gerecht wird.

Das Entwurfs- und Ingenieurkonzept gliedert sich in die Bereiche „digitale Konzeption“ und „Fertigungskonzeption“.

In der „digitalen Konzeption“ wird CAD-Software eingesetzt, um die komplexe Geometrie des Projekts parametrisch zu definieren und alle digitalen Informationen zu generieren, die zur Steuerung des Fertigungsprozesses benötigt werden. Durch die digitale Konzeption werden die zugrundeliegenden Primärregeln der komplexen Fassadenkonstruktion definiert, z.B. Verbindungstypen und Fügungen eines Fassadensystems.

Für die „Fertigungskonzeption“ werden Technologien der Additive Herstellung (umgangssprachlich: 3D-Druck) als eine der möglichen, verfügbaren, computergestützten Fertigungstechniken eingesetzt. Die additive Herstellung wird daher als Prozess untersucht und getestet, um seine Eignung für die Fassadenindustrie zu überprüfen.

Schlüsselworte: *komplexe Fassaden und Gebäudehüllen, parametrisches Design, additive Fertigung (3D drucken), parametrisches Planungskonzept, parametrischer Knoten, N-AM, Li3-Methode.*

Contents

List of symbols

AM	Additive Manufacturing
BESO	Bi-directional Evolutionary Structural Optimization
BET-Method	Absorption method from S. Brunauer, P.H. Emmett, and E. Teller
BIM	Building information modeling
DFMA	Design for manufacturing and assembly
DIN	Deutsches Institut für Normung
EA	Evolutionary Algorithms
EBM	Electron Beam Melting
ESO	Evolutionary Structural Optimization
GA	Genetic Algorithm
$\mathrm{k_B}$	Boltzmann constant
MPB	Melting Pool Boundary
PLA	Polylactide
PREP	Plasma Rotating Electrode Process
R_G	Universal Gas Constant
REP	Rotating Electrode Process
RR	Rejection Ratio
SIMP	Solid Isotropic Material with Penalization
SLM	Selective Laser Melting
SLS	Selective Laser Sintering
STL	Stereolithography
TO	Topology Optimization
vM	Von-Mises

1 Introduction

This dissertation is motivated by the need to create a new method for the design and engineering of complex façades and building envelopes. Complex façades and building envelopes in this dissertation are only dedicated to single layer envelopes in their three main terminologies: stick systems, unitized systems, and free-form shells. The complexity is defined by irregular structures and diverse elements within the same structure. This new method will utilize file-to-manufactory concept, which is the seamless merge of the design process into fabrication. The file-to-manufactory concept will explore the opportunities of parametric approaches and additive manufacturing. These two are growing trends and offer new possibilities. Parametric approaches change how planners/designers perceive building details. Therefore, a new design and engineering concept is needed to tolerate the complexity of architectural geometries. This complexity is increasing, due to the rapid developments in all related aspects; such as façade systems, modeling software, and digital manufacturing techniques. Accordingly, traditional project management is inadequate. The file-to-manufactory concept is a contemporary way to enhance architectural possibilities and maximize efficiency. This concept needs the exploration of specific BIM (Building Information Modelling) and additive manufacturing techniques. This dissertation will explore the requisite steps and will present them as a design and engineering concept that will suit most of the complex façades and building envelopes, and will complement the traditional design and engineering processes.

1.1 Background

1.1.1 Parametric Revolution

Drawings express architectural designs. The drawing medium is the primary constraint for drawing acts and the organization of ideas, spaces, functions, and so forth. CAD software introduces a new definition for the drawing medium, which is the direct translation of drawings within the digital realm. Those drawings are still just a digital representation (code) of the main ideas. The code creates geometry as a result, and this is Direct Modeling, which presents the additive logic of traditional drawings (Figure 1.1).

Luigi Moretti firstly introduced Parametric Modeling in 1939. His unrealized models of a soccer stadium, where the calculated pseudo isocurves generated a final shape with optimized views from every position in the stadium, are the first representations of associative logic and form-finding.

A. Mohsen, *Design to Manufacture of Complex Building Envelopes*,
Mechanik, Werkstoffe und Konstruktion im Bauwesen 56,
https://doi.org/10.1007/978-3-658-30204-7_1

Moretti stated: „The parameters and their interrelationships become the code of the new architectural language, the structure in the original sense of the word. The setting of parameters and their relationship must be supported by the techniques and tools offered by the most current sciences, in particular by logic, mathematics, and computers. Computers give the possibility to express parameters and their relations through a set of (self-correcting) routines" (Mulazzani, 2000).

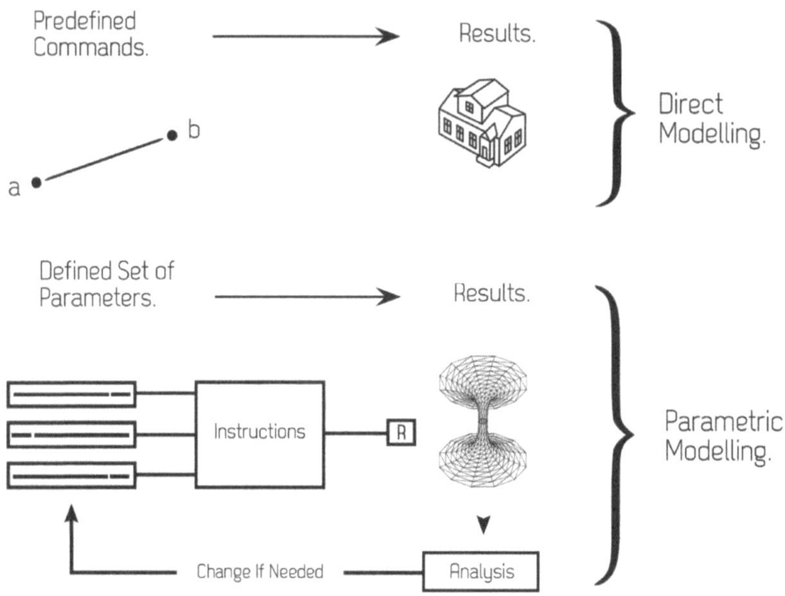

Figure 1.1 The difference between Direct Modelling and Parametric Modelling.

In a parametric medium, CAD is no longer a digital representation of paper and pen. The designer becomes familiar with generating geometries through parameters. These geometries introduce complexity beyond human capabilities. This complexity is driven by how algorithm modeling *–parametric modeling* works. An algorithm needs a question to provide a solution. Equations and parameters define the question (Tedeschi, 2014).

The development of visual software is focused on making scripting more user-friendly because it allows designers with little to no knowledge of programming to utilize it, for example Generative Components by Bentley Systems and Grasshopper by Robert McNeel & Associates. Associative logic can be constructed in this software via the art of nodal diagrams. This constructed logic allows the designer to interact and manipulate digital objects. The manipulated objects introduce a new level of complexity to the field of architectural geometries. This complexity introduces two challenges to the architectural world; the limitation of manufacturing methods, and the intensity of planning time. Therefore the current manufacturing techniques and planning methods need to evolve (Tedeschi, 2014).

The practice and engineering of complex façades/buildings envelopes start with a quick market evaluation. This evaluation explores the products - such as façade systems, window systems, and so forth - that are available on the market, its suitability for the project, and if it can fulfill all detailing scenarios introduced by the level of complexity of the project. Most of the time, it is clear from the beginning that any given project needs a whole new system to be realized. Accordingly, planning takes an entirely new turn, and this turn affects the number of working hours. The requirements needed to design a new system from scratch are the primary factor for that turn and take a lot of time and effort to be fulfilled (Williams, 1999).

On the one hand, traditional manufacturing techniques of façade systems limit the realization of complex façades/buildings envelopes as well as becoming costly. On the other hand, computer-aided production techniques offer new solutions/options to the construction spectrum, and they overcome challenges that face traditional manufacturing techniques.

1.1.2 Additive manufacturing

Additive manufacturing or 3D printing is one of the computer-aided production techniques that has started to grow in the last couple of years, and it offers new solutions to several problems in the manufacturing world. Additive manufacturing is based on the principle that any object can be synthesized through a breaking down into multiple layers. These layers need to be laid at the top of each other to form the final object. Metal-Based additive manufacturing has grown significantly in the last couple of years because of the advantages that technology offers. Terry Wholer, commenting on the formnext fair held in November 2016, said "it was the most elaborate and striking display of AM products and services ever" (Wohlers, 2016).

The 3D printing process consists of several steps. These steps are as follows:

a) Creating a digital model that defines the geometry's external surface.
b) Exporting the targeted model in STL (stereolithography) format; this is suitable for nearly all AM technologies.
c) Importing the STL file to the 3D printer's software - if provided - considering that each 3D printer may have its way to manipulate digital data and generate the appropriate G-code that is needed to run the 3D printer.
d) Printing starts after receiving the data generated by the step before.

e) Creating the final model, which in some instances will require post-processing to finalize and prepare for end-use.

In additive manufacturing, the digital model is generated using visual software; later, it is directly fabricated with minimal process planning. The process planning needs careful and detailed analysis to determine several factors, such as the fabrication order of different features and the usage of tools and processes. In contrast, AM needs a basic under-standing of dimensional detailing, AM machines, and materials. AM processes allow the creation of complex structures and hollow objects that are hard to manufacture using traditional techniques. It offers the advantage of building light constructions in the fastest way possible (Gibson, Rosen, & Stucker, 2010).

Selective laser melting (SLM) is one of the Additive manufacturing processes for metals. SLM is an efficient technology to manufacture metal parts that could be made out of various alloys like titanium, aluminum, and steel. This efficiency comes from the additive manufacturing method, which means that there is no waste material in the process. The process is preferred for the manufacture of the light components, and grid constructions to reduce the mass of the parts. SLM yet faces many challenges. One of these challenges is the physical interactions that happen through the printing process. These interactions develop non-uniform phenomena that occur because of the high number of control parameters. Thermal gradient and thermal stresses are the primary interactions that can develop in the produced element and may affect its mechanical properties. The different sizes of the powder particles and the low construction rate of the process increase the surface roughness of the 3D printed objects. Accordingly, post-processing may be needed based on the purpose of the object given that there are few material alloys suitable for this process. (Gibson et al., 2010).

Parametric modeling increases the complexity of architectural geometries. However, it makes individualized solutions more reliable. For standard manufacturing techniques, individualized solutions are costly and not preferred. On the contrary, AM is not affected by the customization of solutions. The design and engineering of complex façades/building envelopes are forced to individualized solutions most of the time. These solutions go through a long process of detailing and certifying. The merge of parametric modeling and additive manufacturing in the façade industry – from the author's perspective – have good potential to improve the design and engineering process of complex façades/envelopes.

1.2 Motivation

The trend in architectural design and façade technology is an increase in complexity. Baccarini stated in his review in the International Journal of Project Management, "Construction projects are invariably complex, and since World War II have become progressively more so" (Baccarini, 1996). Complex projects demand special treatment, therefore traditional project management becomes inadequate. Hough claimed, "complex projects demand an exceptional level of management, and that the application of conventional systems developed for ordinary projects have been found to be inappropriate for complex projects" (Morris & Hough, 1987).

Project complexity has no precise definition, though different papers have discussed possible definitions. It is concluded that project complexity is defined by two dimensions, each of which have two sub-dimensions (Figure 1.2) (Williams, 1999).

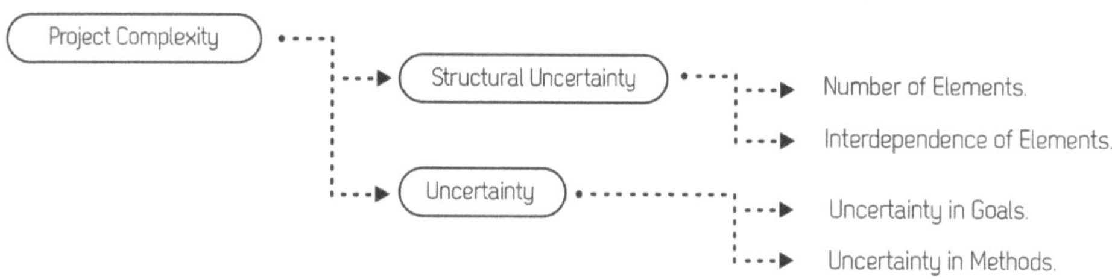

Figure 1.2 The dimensions of project complexity.

State-of-the-art projects have a higher degree of uncertainty than traditional projects, be-cause a similar scope of background knowledge is not available. Therefore, the classical project management techniques cannot handle that level of uncertainty. New methods and ways are needed to design and manufacture modern and complex projects (Williams, 1999).

The complications that occur in complex projects through the design and engineering phases are sometimes unpredictable. The merge between the two previously mentioned technologies introduces a parametric design and engineering method that is no longer project-based. The process could be generalized and applied to most of the complex façades/building envelopes. This façade design and engineering method will introduce new factors into the traditional processes.

The conflicts within any project rise with the increase of the project complexity (Jones & Deckro, 1993). These conflicts are mainly because of the separation between conceptual design and detailed design. Mass production affected this separation even further, which is expressed by the gaps between both design phase and production phase, and designers and factory engineers. The detailed design is usually done after the conceptual design. For complex façades/building envelopes, this leads to an extensive planning process and increases the number of conflicts that arise during the project.

The way an architectural project is planned is not dependent on its complexity, which leads to certain complications. These complications are present directly after the decision to custom design a solution for a project, which expresses a long design process and series of testing and certifying. As a result, a project-based solution is designed. Project-based solutions are suitable only for their exact case and are not manufactured as a system series.

Parametric modeled and digitally fabricated projects can be seen as successful examples of the power of the merge between two approaches, but they are project-based solutions. However, through these projects, the rules for geometric and structural elements were present at the very beginning in the parametric model. The presence of these rules is reflected in the detail design phase and used for the fabrication of the elements.

With parametric planning and digital fabrication processes, a new relation between design and production has been established. This relation is powered by how parametric planning works, where the geometry is defined through an algorithm. This algorithm contains the theoretical expression of the building with a combination of the detailed requirements. Then, digital fabrication processes handle complex detailing with maximum precision and manufacture complex custom parts. This relationship means that the para-metrical model considers the manufacturing process with its limitations; moreover, it can optimize the generated solutions accordingly. This combination allows a blend of form and function and synergy between detailed and conceptual design.

Architectural detailing describes a building in a theoretical way, and defines its technical character. The typical 2D-drawings are the way to represent details, but with the increase of complexity, 2D-drawings are no longer sufficient. Fabrication instructions, assembly instructions, as well as .stl-files and g-codes are now part of the detailing package. In parametric planning, detailing could be scripted and automated, which grants the needed information for digital fabrication.

The file-to-manufactory process describes the merge between parametric planning and digital fabrication. Through this process, the knowledge of the applied production processes, is reflected in the construction details. It was essential to apply these planning concepts in teachings and seminars to examine the process and to develop its sequence. The goal of the *Façade technology* 1 seminar was to plan a small project digitally and to optimize the project according to the available conditions. For the façade industry, file-to-manufacture offers several advantages. These advantages include reduced planning time and fewer manufacturing steps. However, the file-to-manufacture process increases the complexity of the planning process to a certain level based on the use of a new method in the planning process. Therefore, designers/façade engineers have to improve their skills to cope with the change.

New software helped designers to simplify the generation of free-form surfaces. These surfaces then are transformed into free-form envelopes by the façade engineers. These envelopes are broken down into transportable elements in the art of segmentation. This segmentation is the most straightforward translation of surfaces into straight bars, connecting nodes, and (planar) surfaces. The idea is to parametrically transform any complex envelope to its essential elements, then extract the digital information to manufacture the customized elements additively. The idea formulates a new design and engineering concept. This concept will offer several enhancements to traditional processes.

It starts by merging between different planning phases at the early beginning of the design phase (Figure 1.3). Once the geometry of the building is defined, the segmentation process starts to transform the geometry into bars, nodes, and surfaces. Furthermore, the primary requirements from architects, façade planners, and manufacturers, are implemented in the parametric model. The optimization process starts to process these requirements and to consolidate them. Next, the final solution presents the needed manufacturing information, which is a data-set that describes the standard parts and the digital files for the 3D printing process. After the preparation of all parts, the installation process can begin.

This planning process has an impact at all levels of the construction process. This impact starts by merging the design, detailing, and shop drawing phases and runs them in parallel during the project. This merge leads to a decrease in working hours on the one hand, and an increase in the complexity of the process on the other. Another impact is an opportunity to consider the manufacturing process and its logistics, which can minimize corrections and conflicts.

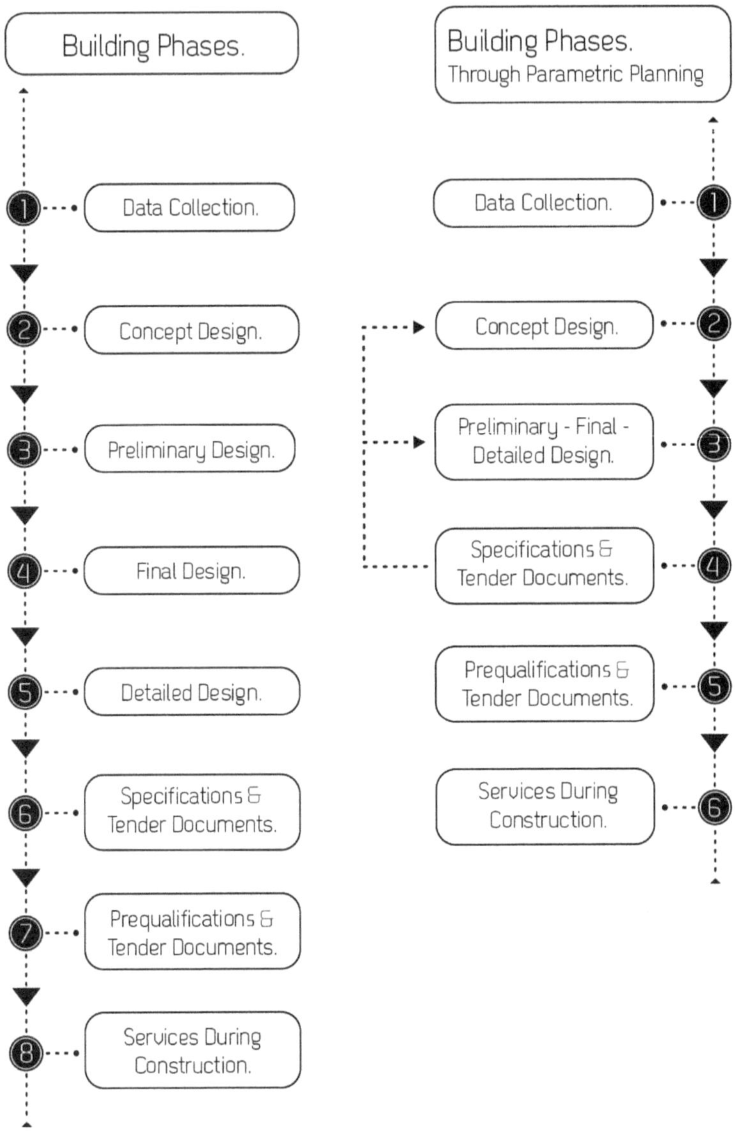

Figure 1.3 Traditional Building Phases vs. Parametrical Building Phases.

1.3 Defining the Problem

Façades/building construction is a project-based action that is defined and controlled with several factors, such as uncertainty, cost estimation, production and management planning. The final costs are usually only known in the late phase of the project. Such a scenario leads to the acceptance of budget escalations, delayed completion, and/or design changes (Jørgensen, 2005).

The cost of construction is usually set at the beginning of the project. The architect and the engineer try to define their solution within the available resources. This burdens the design process and reduces innovation. Standard systems provide usually the most straightforward solution when it comes to the Price-Value relationship. However, standard systems themselves are only applicable within specific ranges and limitations that are defined by their design and specifications.

Customized solutions are always complicated at their cost estimations. The variables and uncertainties are high. However, they offer a functioning solution to innovative, elaborate facades/building constructions. They allow the architect and the engineer to broaden their vision and make something unique.

Value engineering is a powerful approach, but the term has a negative reputation as the activities that aim to cut costs. When it is implemented at early stage of a project, it will help design to target cost (Jørgensen, 2005). Value engineering as a topic is not investigated in this dissertation. However, all design decisions were met with the tendency to value engineering principles.

Complex façades/building structures tend to be complicated and costly, which limits the possibilities to realize more sophisticated buildings. A planning method is needed to structure the merge between parametric design, additive manufacturing, value engineering, and standard façade systems. This merge leads to a new way to design and engineer economic complex façades/building structures.

The endless passion of architects for individual, unique designs is always growing. Accordingly, the tendency for individualized solutions increases. The parametric planning concept will accommodate complexity and individualization. This compromise structures the steps of the process.

The manufacturing limitations such as maximum profile length and maximum manufacturable surface area affect the main parameters of the generated grid. The available crosssections and connecting components define the structural behavior of the building envelope.

Material properties of 3D printed material are available related to its print direction. The 3D printed element will contain different print orientation within its volume. The more general approach needs to be investigated to reduce the computing time of the stress analysis of 3D printed metal parts.

The prediction of 3D printed materials is still challenging because of the anisotropic behavior. To optimally utilize topology optimization for an element, the material properties have to be carefully defined.

The additive manufacturing process is not regulated in the building industry. Accordingly, the generated objects are treated as custom parts and need to be certified.

1.4 Research Questions

The following question will form the matter of the investigation for this dissertation:

How to use parametric design and additive manufacture to create Complex Skeleton Metal Envelopes?

The answer to this question consists of several statements. Each statement explains a set of facts or principles that this dissertation will generate and prove. Therefore, the following sub-questions will guide the scientific approach to formulate and support the target research.

Chapter 2:

- What is the state of façade technology?
- What are the technical possibilities of façade/nodal systems currently available?
- What are the limitations of the current technologies?

Chapter 3:

- What are the basics of additive manufacturing of metal alloys?
- What are the controlling factors for the selective laser melting (SLM) process?
- How can the behavior of a 3D printed metal elements for building industry be predicted?

Chapter 4:

- What are the possibilities for translating complex envelopes into parametric models?
- What are the defining factors of the visual presentations of façade elements?
- What are the façade functions and their effect on the parametric design?
- What are the design guidelines that ensure manufacturability of AM elements?
- What other factors need to be investigated to regulate the efficiency of the process and its results?

Chapter 5:

- How do the parametric planning concept develop as a whole?
- How the experimental stages evolve through the study?
- What is the evaluation of the results?

This dissertation is oriented to the façade technology, which means that the sub-questions will be answered in favor of the relation between the involved parties, such as architects, façade planners and manufacturers; and to the design and engineering process of complex architectural projects.

1.5 Approach and methodology

The study of different sectors involved will follow an engineering approach based on scientific principles. The engineering design process matches the objective of the dissertation. This objective defines a functioning planning system for complex envelopes. There are several parts of the system that are not currently regulated by the building industry. Therefore, the requirements of both the design and manufacturing of custom elements have to be defined through testing and prototyping. These elements will be evaluated and subject to modifications until the solution meets the requirements or proposes future developments.

The engineering design process will be followed to answer the central research question (Figure 1.4). This serves to define the main factors for the design and engineering of complex envelopes. These factors will help to determine the scope of work, and to define the needed studies for each phase of the process.

Chapter 1 includes introductory information. This part consists of the background of the topic, the definition of the problem, the motivation, the research questions, and the scientific approach. Chapter 2 introduces the state-of-the-art of façade technology, where the basics of skeleton façade systems are introduced, as well as the different nodal systems. Several recent case studies are analyzed. Finally, the limitations of the current systems and the objectives of the dissertation are declared.

The basics of additive manufacturing and AM of metal alloys are in Chapter 3. Different factors that affect the process are suggested. A method for testing steel additive manufactured parts will be studied and performed. In Chapter 4, the geometrical specifications for segmentation of complex geometries will be analyzed. The digital representation of the connecting points is studied to define the software limitations that may occur. Several aspects, such as façade functions, design for manufacturing, topology optimization, and cost analysis, are analyzed to define their effect on the planning cycle. The collected results help the developments that are listed in Chapter 5.

The solutions/developments are presented in Chapter 5. The knowledge gathered from the previous chapters is utilized to define a final solution. Developed solutions are discussed and evaluated. Finally, in Chapter 6, the conclusion of the research is declared, and further options for future developments are suggested.

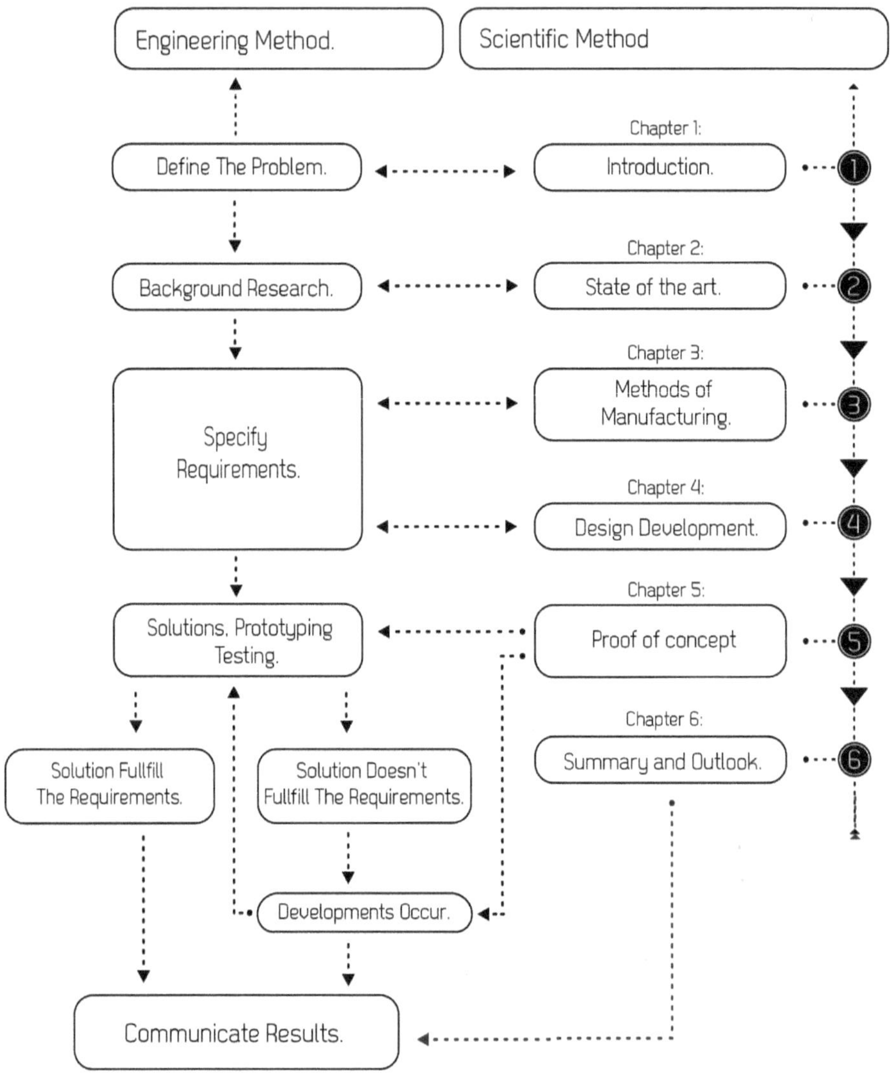

Figure 1.4 The relation between the Engineering Steps and the Scientific Method.

To structure the underlying principles and rules for the dissertation, a draft of a planning concept is proposed. This planning method is called "Li3-Method", and it consists of three parts (Figure 1.5). These parts are: (1) the digital conception of the process, where the realization of the structural elements is generated. (2) the manufacturing conception of the individualized elements, where 3D printing of metals is investigated and the elements are engineered. (3) the assembly process of the whole structure elements, where predefined solutions are integrated into the parametric model at the early beginning.

Several challenges occur during each phase and within the interaction of certain stages with each other. The sequence of the interactions defines these challenges.

The digital conception starts with the definition of the basic geometry of a building and its grid. Then the transformation into segments is performed. Various factors – such as profiles' orientation and planarity – are examined to prepare the structural model. The engineering of the structural model begins by defining the primary parameters – such as the chosen system and its structural capacity – to dimension the main profile of the structure. The optimization of the whole geometry is done to optimize the design on the cell level (façade element). Once the profile specifications are defined, the connecting nodes are designed based on the required geometrical rules – such as connection design, and node location – and then digitally generated. Specific properties of the metal used for 3D printing of the nodes are considered in the topology optimization process. Specific integrity tests are then run on the nodes before the last generation of the production data.

The manufacturing conception starts with defining the selected 3D printing machine and metal powder. Then the printing parameters are defined, and tensile specimens are tested. Because of the anisotropy behavior of 3D printed objects, more sophisticated tests are done – Validation model, to improve the values that express specific material properties. These values will be used to run the topology optimization process – Topology optimization is used as a factor for cost reductions. The most sophisticated node case is defined, 3D printed, and prepared for testing. These results determine the final modifications before manufacturing.

The assembly process starts with the design of the connections between the different elements of the structure, such as nodes, profiles, and glass or metal panels. This connection affects the engineering of the profiles. Specific rules for logistics affect the generation of the nodes and the choice of the profiles.

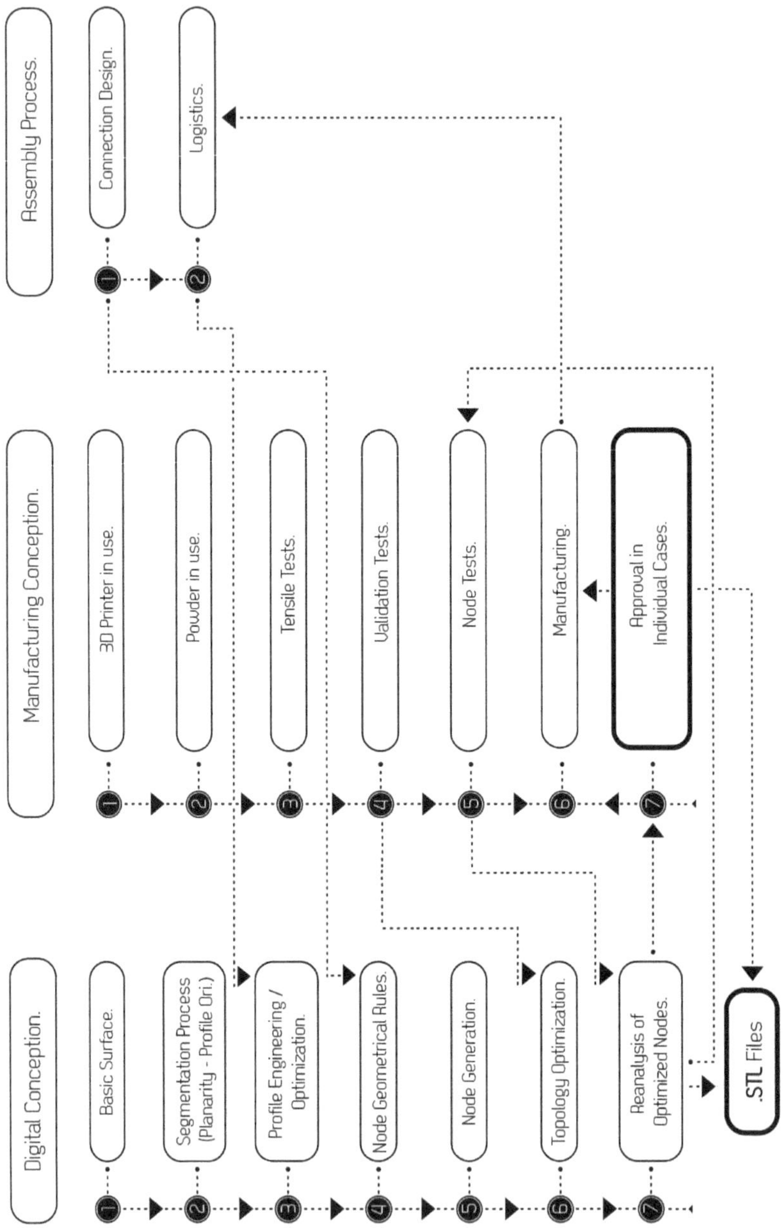

Figure 1.5 The interaction between the different stages at parametric planning phases.

2 State of the Art

2.1 State of the Art

At the beginning of the nineteenth century, typical building constructions were replaced with load-bearing constructions in skeleton form. The skeleton form released the façade constructions of their load-bearing function. Therefore, the development of façade constructions created new possibilities. Schaal, in the Sixties, defined the forming of façade construction into two types: linear and surface forming (Schaal 1961). Linear forming is the stick system (mullion and transom construction), where the façade consists of mullions, transoms, and glass panels or opaque panels. Surface forming uses panelized (unitized) systems, in which the façade element is fabricated in the factory and all other elements – such as windows, doors, and opaque panels – are directly integrated. A third type that developed in modern architecture was a free-form single layer envelope structure. The main components of this type are connecting bars, planar/curved surfaces, and connecting nodes. Free-form and complex-shaped envelopes have several factors – such as optimal design, material strength, connection design, and curvature flow – that have to be appropriately analyzed and designed to be able to construct a reasonable single layer structure (Figure 2.1).

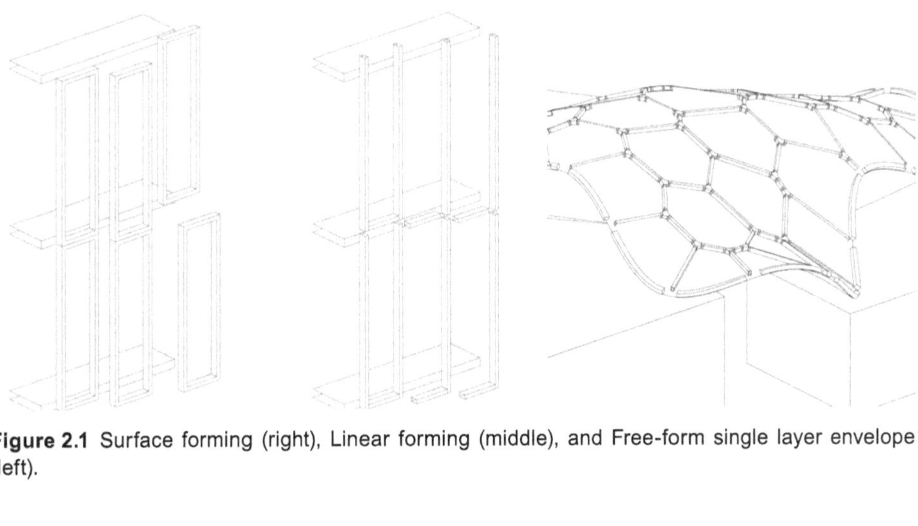

Figure 2.1 Surface forming (right), Linear forming (middle), and Free-form single layer envelope (left).

A. Mohsen, *Design to Manufacture of Complex Building Envelopes*, Mechanik, Werkstoffe und Konstruktion im Bauwesen 56,
https://doi.org/10.1007/978-3-658-30204-7_2

Each type of construction had developed its unique variety of connectors. These connectors developed through time to improve the structural performance of the profiles, and to increase the connection flexibility between the profiles.

For stick systems, the primary connections are mullion-mullion connection, mullion-transom connection, and mullion-slab connection. For unitized systems, the primary connections are unit-unit connection and unit-slab connection, as well the connection between the unit's profiles at the corners. For the free-form single layer envelopes, the connecting nodes are the most significant points of the structure. The nodes can be defined by the angles of its connecting sides. The stiffness of these nodes determines the stability of the whole structure (Figure 2.2).

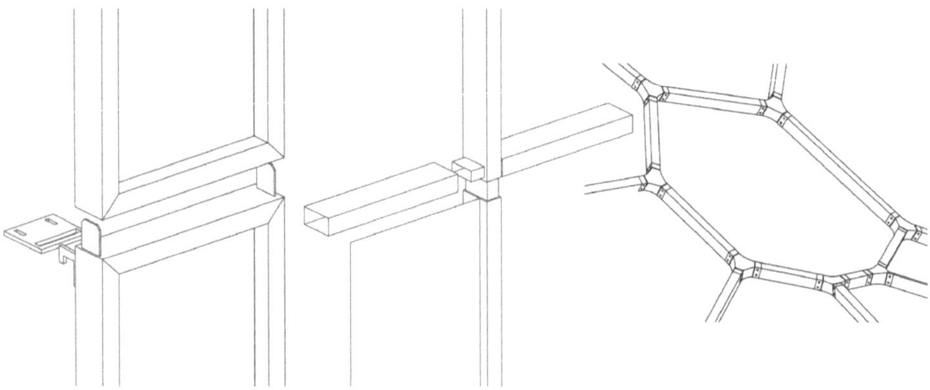

Figure 2.2 Connecting points of Unitized systems (left), Stick systems (middle), and Free-form single layer structure (right).

One of the first glazed wall constructions that formed the first generation of curtain walls is Palacio de Cristal in Madrid, Spain (Figure 2.3). Spanish architect Ricardo Velázquez Bosco designed it after being inspired by the Crystal Palace in London in 1887. Both were constructed out of iron and glass, which were used to form the framework of the buildings. They were the first generation of economic façades to be built with cast iron construction elements. These elements were designed in a way that allows them to be dismantled if needed. This allows the possibility of moving the whole structure and re-constructing it on another site (Renckens 1996). Therefore, most of the connections be-tween the elements were bolted connections to simplify the disassembly process.

Figure 2.3 Crystal Palace – Madrid, Spain.

Later in 1905, the Antwerp railway station was designed by Louis Delacenerie (Figure 2.4). It is considered one of the finest train constructions in Belgium. The railway station roof covering consists of a glazed roof and arches, composed of cast iron profiles that were erected and glazed. These roofs allowed the buildings to be full of light, which increased the trend towards more slender metal/glass constructions from the middle of the nineteenth century (Renckens 1996).

Figure 2.4 Antwerp Main train station – glass roof.

The curtain wall came as the new construction method for walls and roofs. This construction consists of iron and glass. The trend to allow more light inside buildings, encouraged architects and engineers to improve the construction of curtain wall systems. In the Twenties, architects started to utilize curtain wall systems for buildings with two or more stories. The workshop building of the Bauhaus in Dessau (1925-1926) is an example of this era (Figure 2.5). This kind of façade was the first example of stick systems as they are known nowadays.

Figure 2.5 The Workshop Building- Bauhaus, Dessau, Germany.

At the beginning of the Fifties, new technologies such as air conditioning, artificial lighting, industrial glass production, synthetic rubber production, and prefabrication of building components offered further developments for curtain wall systems. These systems frequently used fabricated steel, but there are some examples where brass profiles were used. Ludwig Mies van der Rohe designed the S.R. Crown Hall (Figure 2.6), which was completed in 1956. Mies van der Rohe presented a new image in basic steel and glass construction.

Figure 2.6 S. R. Crown Hall designed by Ludwig Mies van der Rohe.

World War II limited the use of aluminum in the façade industry. Later, however, aluminum had a better price-performance ratio than steel. In the seventies, almost all façade profiles were fabricated out of aluminum. John Heinrich and George Schipporeit, which were students of Ludwig Mies van der Rohe, were the designers of the iconic Lake Point Tower (Figure 2.7). This tower completed in 1968 and it was the tallest apartment building in the world at that time. The building had a bronze-aluminum curtain wall and double-glazed windows.

Figure 2.7 Lake Point Tower by John Heinrich and George Schipporeit.

The energy crisis in 1973 influenced the façade industry by determining the need for thermal insulation in standard façade construction. At this point, the development of thermal brake systems for the aluminum profiles started, and so began the second generation of aluminum curtain wall systems. During the Eighties, the technical development of façade constructions took a step back, and sales/maintenance became more feasible. Through the Nineties, the increase of industrialization brought the tendency to build higher, quicker, and lighter. Accordingly, element façade systems gained more importance than stick systems (Renckens 1996).

The term glass façade usually refers to the separation between inside and outside. Building envelopes influence many aspects of the building - such as view, ventilation, building services, and possible loadbearing functions (Knaack, Klein et al. 2014).

The execution of glass façades is subject to the change of related technologies, such as profile extrusion and glass panel manufacturing. Therefore, execution methods are consistently developing. As a consequence of economic growth and automation, standard systems have become more favorable. However, the rapid development of complex geometries is challenging standard systems and is increasing the complexity of planning processes.

2.2 Recent Case Studies

Three projects were selected to demonstrate how the most recent solutions to the engineering of complex envelopes. The selection criteria are Linear forming - customized stick system (New Munch Museum), Panelized/surface forming - customized unitized system (Mainzero), and customized single layer gridshell structure (Hangar 8). Besides, the three case studies are customized solutions, built under several European regulations, and of a unique character.

2.2.1 New Munch Museum

Herros Arquitectos won the first place in the competition held for the design of the New Munch Museum (Figure 2.8). The project is about 12 300 m2 floor area and contains both recreational areas and museum exhibit areas. Bollinger and Grohmann Engineers were assigned to design the 7000 m2 glazed façade. The façade must fulfill passive houses requirements as well as form a single ventilated skin with an exterior layer made out of perforated aluminum panels.

Figure 2.8 Munch Museum – Bollinger und Grohmann ©.

Due to the panels' huge size and high-security requirements, the utilization of standard systems was not possible, and a customized façade system was developed. The façade system designed is a stick system. Accordingly, the different façade components were transported to the building site and assembled on site (Figure 2.9). The customized profiles were designed with a special integrated system of connectors. A steel structure was designed to support the second perforated aluminum skin. This structure has to pass through certain components of the first curtain wall skin to be anchored to the main structure.

The planning process was intensive and time consuming, due to the increased number of hours that were needed to design and engineer the customized system with all its features. The second skin was iteratively simulated with different panel thicknesses and supporting structures, due to high wind speed at the harbor.

Figure 2.9 Munch Museum construction process– Bollinger und Grohmann ©.

The inner façade layer consists of insulated profiles that hold the glass panels. The outer layer consists of perforated aluminum curved panels that are supported by its frame. The aluminum panels are hung on I beams that extend all along the façade. The I beam is a second layer construction that accommodates a grating platform in-between (Figure 2.10).

Figure 2.10 Munch Museum Mock-up (second skin connection to the main façade) – Bollinger und Grohmann ©.

2.2.2 Mainzero

In the city center of Frankfurt, Germany is Mainzero, which consists of three basement floors and nine floors (Figure 2.11). The project has around 17 500 m2 façade surface, where some parts were renovated, and others were newly constructed. The architect was KSP Jürgen Engel Architekten, and Bollinger and Grohmann Engineers were assigned for the structural engineering and the façade planning.

Figure 2.11 Mainzero – Bollinger und Grohmann ©.

Due to the panels' huge size and high-security requirements, a customized façade system was developed. The façade system is a unitized system. Accordingly, the prefabricated façade elements were transported to the building site and mounted there (Figure 2.12). The part of the façade that consists of existing natural stone was renovated. The system went through several changes to obtain the requested visual appearance. The huge size of the elements requested an intensive series of engineering and testing processes.

Figure 2.12 Façade assembly of Mainzero – Bollinger und Grohmann ©.

The custom designed façade system accommodates a triple glazed glass panel. The same panel has profiles with different depth to match the requested design. One of the side profiles has integrated ventilation flap. The panels are connected on the sides through an interlock system; and they are connected with a fixed point at the top, as well as a movable point at the bottom of each element to absorb the building tolerances (Figure 2.13).

Figure 2.13 Mainzero, façade system – Bollinger und Grohmann ©.

2.2.3 Hangar 8

Hangar 8, in the airport of Salzburg, Austria, is one of the two hangars that belongs to the Red Bull Company (Figure 2.14). Hangar 8 is used for the restoration and maintenance of the aircraft that are displayed in Hangar 7, which is located exactly in front of it. Architect Atelier Volkmar Burgstaller of Salzburg, Austria, designed Hangar 8 in the form of a quarter of an ellipsoid with a width of 64 meters and a length of 58 meters. Bollinger and Grohmann Engineers were assigned for the structural planning of the project

Figure 2.14 Hangar 8 – Bollinger und Grohmann ©.

The hangar is designed as a lattice gridshell construction, and it has significantly small steel/square meters ratio. The gridshell consists of parallel primary frames that are drip welded to round pipes. The round pipes are superposed to form the triangular grid (Figure 2.15). The structure was custom designed for the project to create the designed form.

Figure 2.15 Hangar 8 (gridshell assembly process) – Bollinger und Grohmann ©.

The node point was formed through the connection between the structural elements. The cladding panels were placed on L shaped brackets. The brackets have long holes to absorb the tolerances that may occur through the assembly (Figure 2.16).

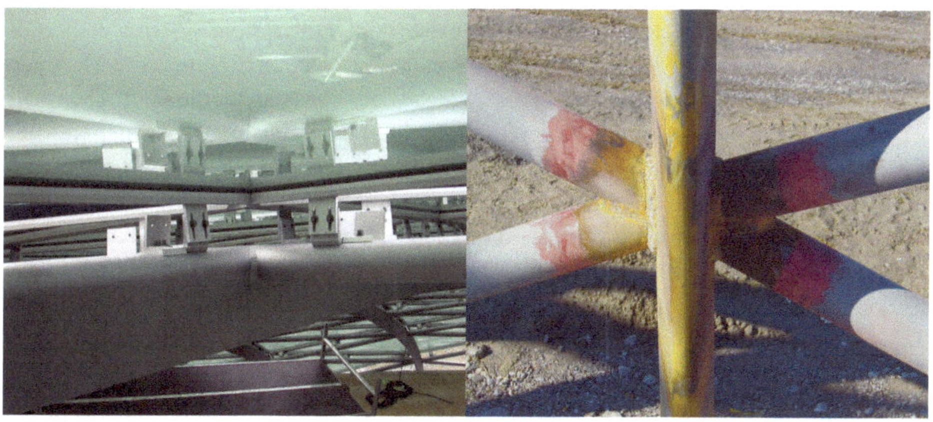

Figure 2.16 Hangar 8, Node and Bracket connections – Bollinger und Grohmann ©.

2.3 Building envelope systems

2.3.1 Linear forming - Stick Systems

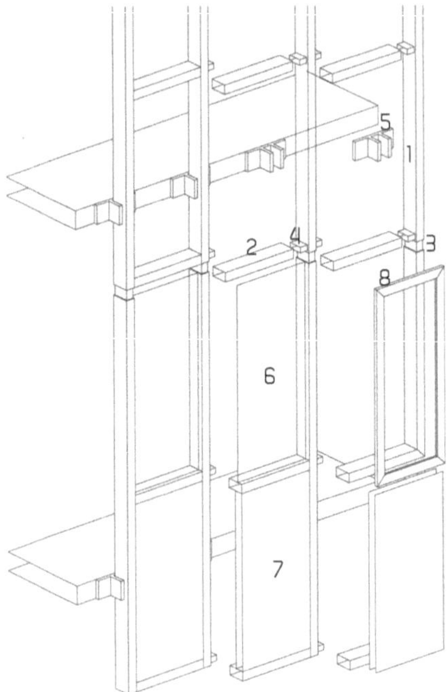

Figure 2.17 Schematic of Stick systems.

Stick systems represent the first generation of curtain walls. These systems consist of different parts that are assembled to form the façade construction (Figure 2.17). The main parts are the (1) vertical profiles (Mullions), which generally span floor to floor. The vertical mullions are fixed to the main structure through (5) the fixing brackets which transfer the loads (self-weight, wind, and snow) to the main structure of the building. The vertical profiles are connected to one another through (3) spigots, where a slip joint is usually used to absorb the building movements. The (2) horizontal profiles (Transoms) are fixed to the vertical profiles through (4) spigots. There are different methods to design this connection, for example: butt connection, butt-jointed connection, plugged connection, and slide-in connection (Figure 2.18).

The horizontal profiles transfer the load of the filling elements to the vertical profiles, then the vertical profiles transfer that loads to the main structure. The filling elements are the components that are fixed into the vertical and horizontal profiles, such as (6) glass panels, (7) spandrel panels, and (8) windows (Mayine 2013).

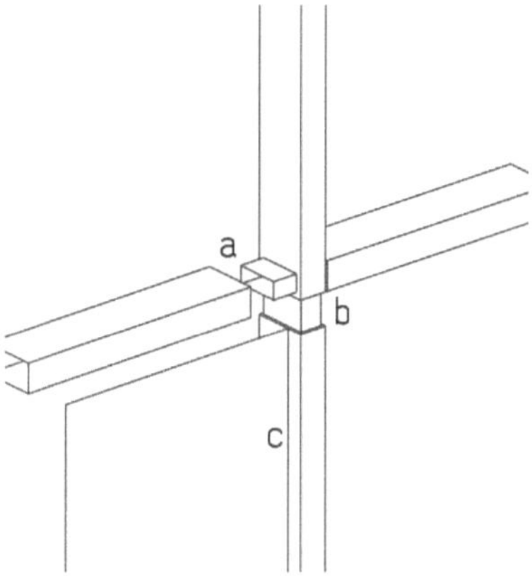

Figure 2.18 Stick system main connection point. a) mullion-transom connection, b) mullion-mullion connection, c) mullion-glass/panel connection.

Usually, stick systems are built from standard components, but they may have different interfaces. These interfaces are custom designed to include different accessories, such as shading systems, ventilation systems, and other building services components. All façade elements are transported individually to the construction site. The façade erection starts with mounting the dimensioned mullions to the building skeleton, fixing the transoms to the mullions, and inserting the filling components to the façade grid (Renckens 1996).

The stick curtain wall systems have to allow for a certain amount of movement. These movements are usually either from wind or building movements. Therefore, each mullion has to contain one sliding connection. The infill elements are mounted with protection blocks to limit their rotation (Mayine 2013).

Stick systems are widely utilized nowadays because of their good availability in the market. They have been developed over time from an economic motivation that results in cost-effective solutions. However, the standardization of the process limits the possible solutions, which makes stick systems unsuitable when the complexity of a desired façade exceeds the system limitations. The main limitation is the angle connection range between the mullion-to-mullion or the mullion-to-transom connections. This angle range defines the system ability to form the envelope design. Because of standardization, the profiles, the glass panels, and the opaque panels have size limitations based on the manufacturing process used.

2.3.2 Panelized/Surface forming - Unitized Systems

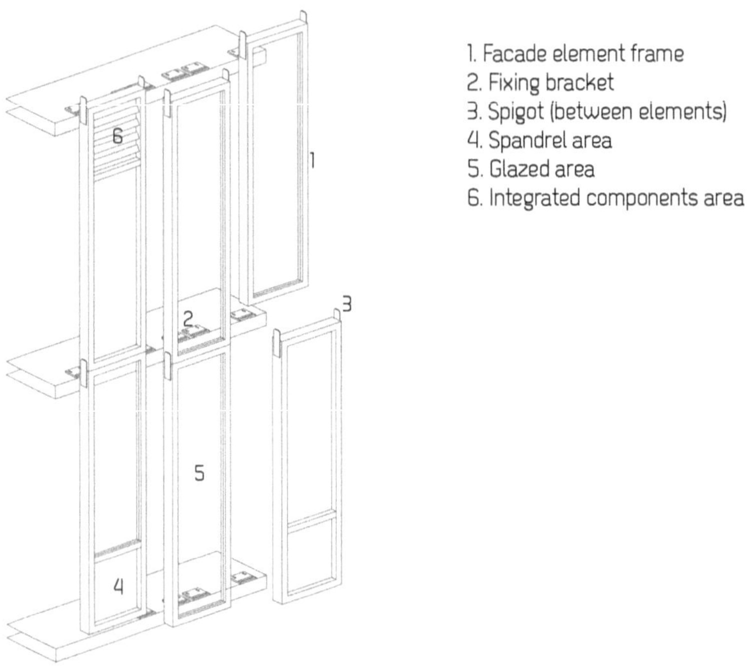

Figure 2.19 Schematic of unitized systems.

Unitized systems came as a result of applying work-saving methods; thus, the complete façade element is prefabricated in the factory. These methods granted precise fabrication, rapid assembly, and low labor capacity. Unitized systems are usually utilized in high-rise buildings, where the elements are highly repetitive and are mounted from floor to floor (Figure 2.19). These elements consist of (1) the façade element frame that contains the designed components such as (4,5&6) spandrel panels, glass panels, and services panels. The elements are usually connected through interlocking systems using (3) spigots and are fixed to the main structure through (2) fixing brackets (Figure 2.20).

Unitized elements are prefabricated elements. These elements are double mullions that interlock with one another. All the required accessories are integrated into the façade elements, then transported to the construction site. At the construction site, the elements are transported with cranes and positioned to their exact locations.

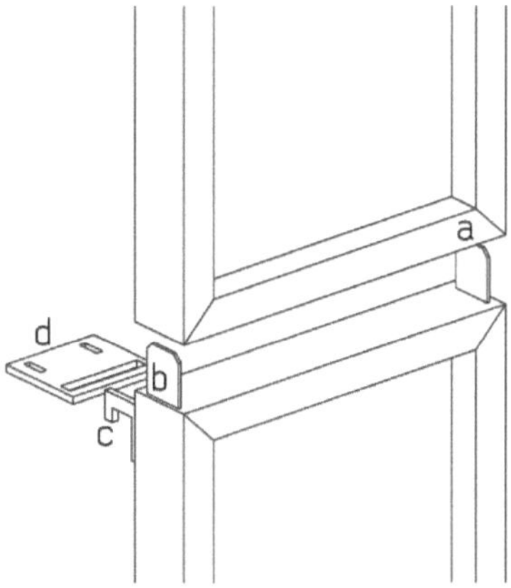

Figure 2.20 Stick system main connection point. a) unit corner connection, b) spigot connection between elements, c) hanging connection, and d) fixing bracket element.

The unitized systems required greater tolerances than stick systems, whereas industrial manufacturing allowed better quality control for the unitized systems to be performed (Renckens 1996).

Unitized systems offer a high-speed construction method that can be highly monitored. However, these systems reach their limits quickly in cases of free-form projects with high degrees of complexity. As a result, a custom-designed system must be developed to satisfy those projects' constraints.

2.3.3 Different nodal systems for free-form single-layered building envelopes

Free-form envelopes in this work are defined as: one-layered construction, individual geometric and structural conditions, and developable surfaces. Therefore, the structure is constructed from straight bars that are fixed at each end with a force-fit node. The use of straight surfaces that are fixed to straight bars, which are connected through compact nodes, is considered for economical desirability (Figure 2.21).

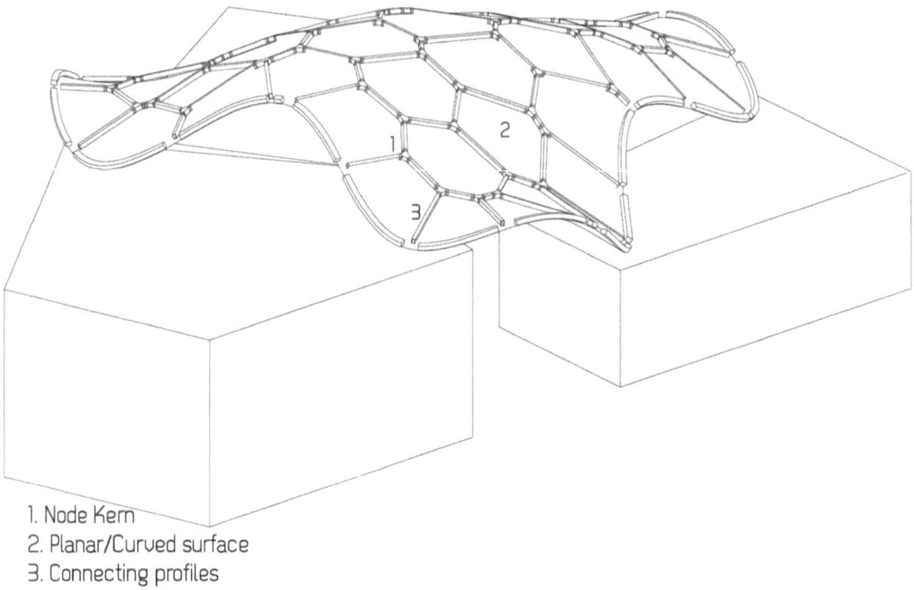

Figure 2.21 Schematic of Free-form single layer structure.

The main idea of industrial construction is the standardization of systems and products in a deliberately arbitrary manner. This aims to the increase of the economic efficiency for the produced elements through the developments of easy-assemble, standardized building parts (Wachsmann 1959, Mengeringhausen 1966).

Parametrization came as the idea to expand standardization of elements. Therefore, elements are parametrically designed, which means that the geometric size and/or the structural performance were constrained to limited parameters that aim towards variable and rapid design.

Free-form envelopes are transformed into a free-form structural network. This structural network is made out of straight bars that have to be connected through structurally designed and force-fitted connections. These connections are the node points, which form the required geometry, ease the orientation and the assembly, of the connecting bars. Arciszweski introduced a structural model that represents the conceptual design of a connecting node. This concept specified the required elements for a structural node (Figure 2.22). These elements are (1) the assembly connection between the node and the connecting bar, (2) the connection between the connecting bar and the bar end, (3) and the connection between the node kern and the node end (Arciszewski 1984, Stahr 2006).

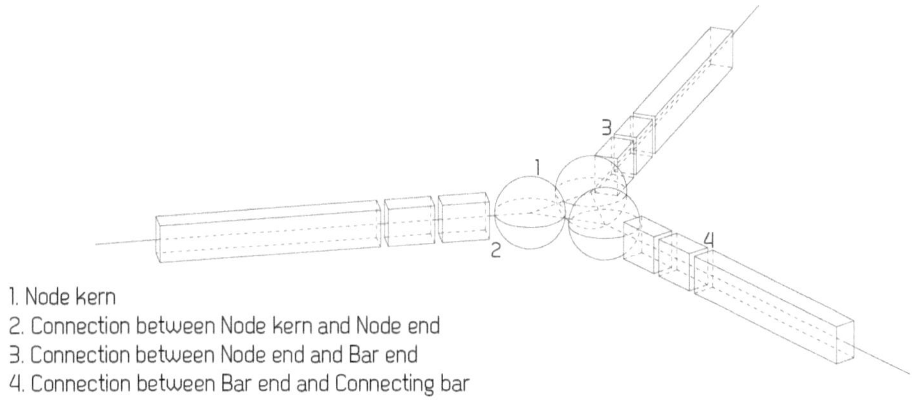

Figure 2.22 Arciszweski Structural node model.

Nodes, bars, and their connecting elements are the basic elements of any spatial structure. These elements fulfill the following specific functions: 1) forming a primary bearing structure, 2) forming a decent boundary area that serves as linear support for the cladding elements, and 3) forming the requested free-form geometry that is intended to be generated (Stahr 2006).

The idea of a universal node is still not achieved, but many interested parties are working on the idea. Nodes are defined through required geometrical conditions and subsequent structural integrity. Nodes currently have several kinds of connection types, such as bolted, welded, or custom-fabricated. The regularity of the structure determines the geometry of the nodes and a correspondingly suitable fabrication technique, such as casting, extrusion, folding, special working, and so forth.

Each building regulation prefers a particular type of node connection, because of the diversity of the testing and certification methods for the systems. Each system offers a certain degree of freedom, which is represented by the range of geometry conditions that the node can accommodate. Still, the need for a custom designed connection is often present, when a free-form project exceeds the system's designed capacity. In the upcoming section, different solutions will be discussed.

Splice node Type

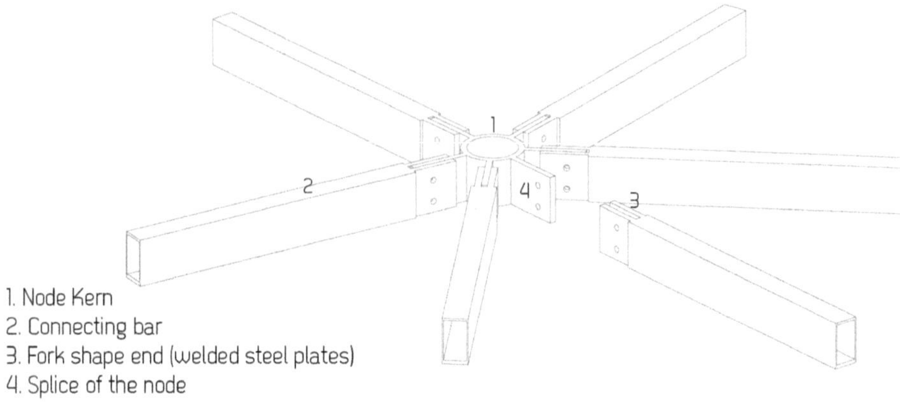

Figure 2.23 Schematic of Splice node type – POLO-1.

The splice node connector POLO-1 was designed by Polonyi & Fink, Cologne, Germany. This node has a cylindrical or prismatic kern that can connect up to six vertical splice plates (Figure 2.23). The connecting end of the bar is fork-like, which connects to the splice plates by two bolts. The splice plates accommodate the twist angles of the connecting bars. Accordingly, higher bending moments can be transferred. This node type was used to construct the canopy roof of the railway station in Cologne (Stephan, Sánchez-Alvarez et al. 2004).

Gartner Type

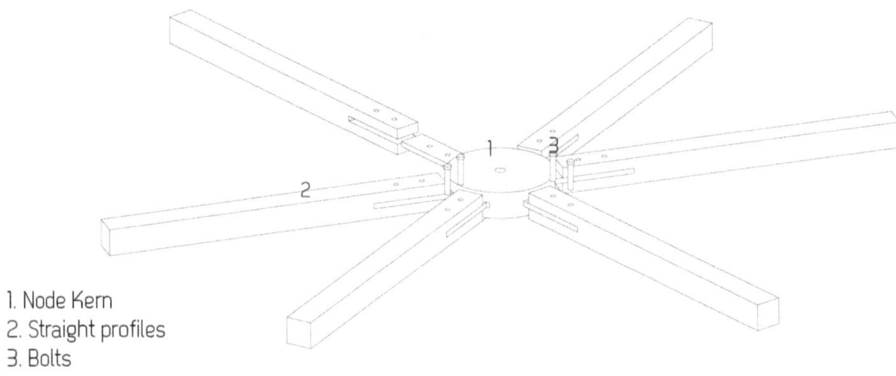

Figure 2.24 Schematic of Node Gartner type.

The star-shape node was designed to absorb normal forces and partially bending moments (Figure 2.24). The geometry of this node grants the formation of triangular grids. Water jet cutting and double milling are used to manufacture it, as the production data is directly transmitted to five axial milling machines. The star-shape node is connected to rectangular bars through bolts. It was used to construct the roof of the headquarters of DZ-Bank, located in Berlin, Germany (Schlaich, Schober et al. 2001).

Waagner-Biro Type

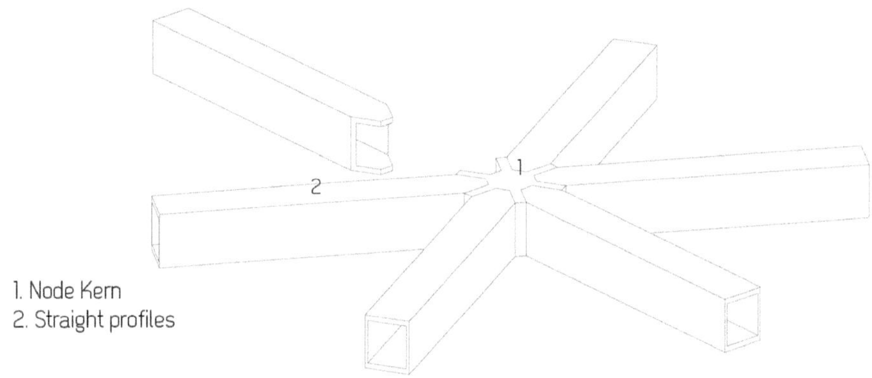

Figure 2.25 Schematic of Node Waagner-Biro type.

This node is built as a massive solid element to absorb bending moments (Figure 2.25). It is manufactured in a one-step process, wherein the node is cut out of a thick plate with smaller height compared to the connected bars to avoid any collision with the glass surface. The connected bars are made out of thick welded plates, which then are welded to the node. It is suitable for slightly bent structures, as the increase in the bend degree of the surface also increases the complexity of node and glass manufacturing. This node form is easy to manufacture, and the welded connections are easier to adjust and construct (Barnes, Dickson et al. 2000).

Seele Type

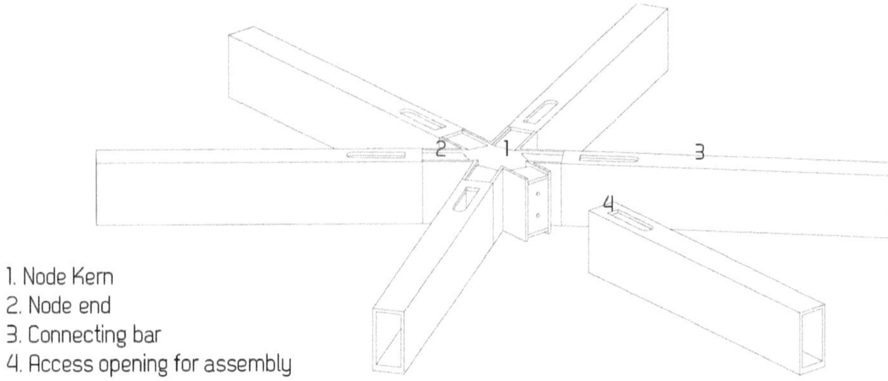

Figure 2.26 Schematic of Node Seele type.

This node consists of a node kern that is a cut out of thick plates, a node end that is assembled out of different cut plates to accommodate the geometrical condition and to provide a clean connection for the bars, and a connection between the node and the bars that is a bolted connection, accessible through integrated access openings (Figure 2.26).

The additive technique provides the node with a wide range of adaptability. Another ad-vantage is its hardly noticeable visual appearance (Behling, Brensing et al. 2010).

Mero Type

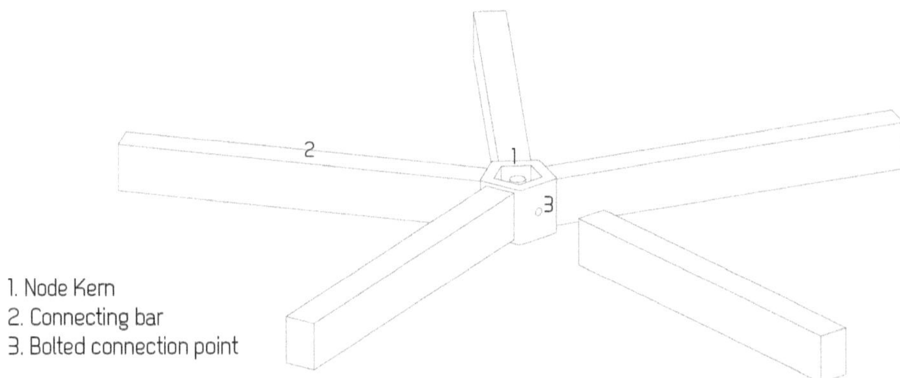

Figure 2.27 Schematic of Node Mero type TK.

Mero nodes were originally designed as a pin connection for double-layered structures. However, the need for single layer structures with cladding elements increased. In accordance, Mero introduced new jointing system called the Mero plus systems (Figure 2.27). This system can construct a variety of curved structures, transfer shear forces, and in certain cases resist torsion and moment forces (Makowski 2002).

The node is planar ring-shaped that can be connected with 5 to 10 rectangular profiles. The node and the bars have the same section depth. Each bar is connected to the node with one bolt. The node transmits the shear and rotation forces, and it is suitable for triangular latticed shells (Lan 1999).

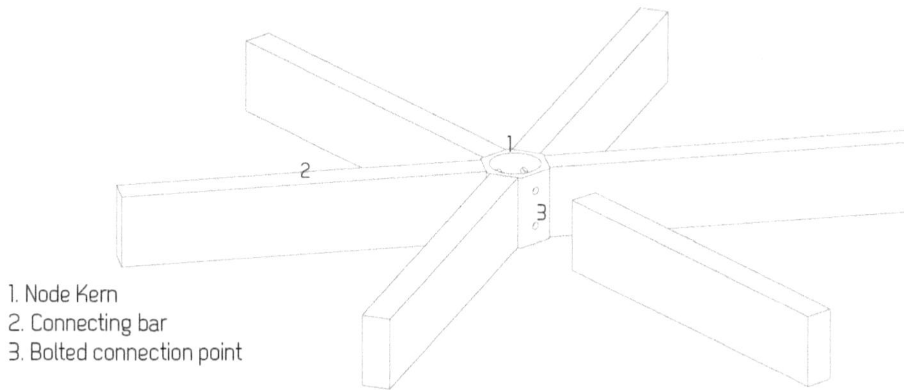

Figure 2.28 Schematic of Node Mero type ZK.

Mero node type ZK is a cylindrical node that can be connected with 5 to 10 rectangular profiles (Figure 2.28). Each bar is connected to the node through a double bolted connection. This node transmits bending moments, and it is suitable for double curved surfaces that consist of trapezoidal cells that need rigid flexural connections (Lan 1999, Makowski 2002).

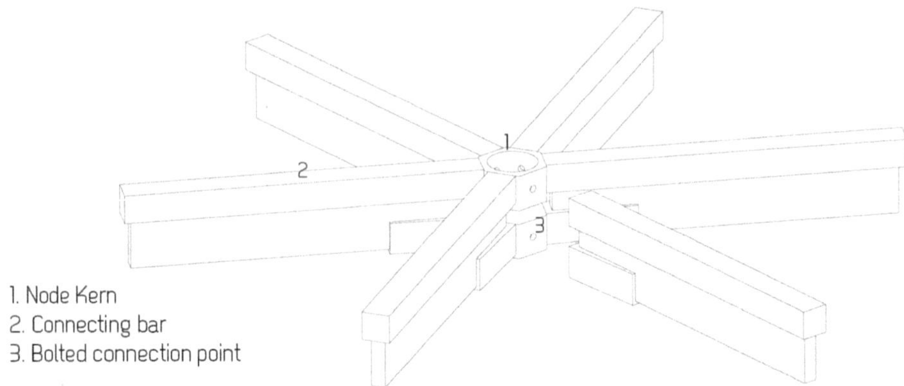

Figure 2.29 Schematic of The Double Node - Mero type.

The double-node is the latest version Mero has presented. The node consists of two dish nodes (Figure 2.29). One is connected with the upper chord and the other with the lower chord. This connection is either bolted or welded. The machine-shaped surfaces of this node adapt to the surface complexity (Hwang 2010).

2.4 Limitations of current systems

For conventional façade systems and spatial structures, the connecting system is the most crucial point of the structure. This connecting system is responsible for the accommodation of the geometry, and the local and global stability of the whole structure. Therefore, the connectors of those systems are complicated, as they are not connected to an equal number of elements in some cases and are positioned in three-dimensional space in other cases. Companies and researchers compete in offering different solutions. It is reported that over 250 different types of nodal connections have been suggested or used in practice, and that around 50 commercial companies worldwide have specialized in manufacturing connecting systems for spatial structures (Lan 1999, Hwang 2010).

With the architectural demands for transparency and the tendency of the architects to distance themselves from simple planar envelopes, single layer envelopes gained more interest. Accordingly, the challenge became the development of connecting nodes that are not subject to only axial forces but also to bending moments, which the usual connectors of the standard façade systems cannot accommodate. Later, free-form shapes and complex envelopes were introduced to the architectural world, which increased the interest in connecting systems. The design/planning process has to be carefully investigated to offer a fully functioning connecting system.

The primary purpose of a façade is the separation between the outer environment and interior spaces. Façade functions can be defined in 5 categories, based on the increasing level of detail. These categories are main function, primary functions, secondary functions, supporting functions, and detailed supporting functions. The first two primary functions are creating a durable construction and allowing reasonable building methods (Figure 2.30).

To create durable construction, the following is required: the structure has to sufficiently bear the subjected loads, the façade has to enable decent management of the water and vapor in the construction, and the materials and components have to be in a working condition for the life cycle of the construction. To allow reasonable building methods, the following is required: the design and management processes have to be refined, the production methods have to be reconsidered, the transport has to be planned, and the assembly methods have to be carefully engineered (Klein 2013).

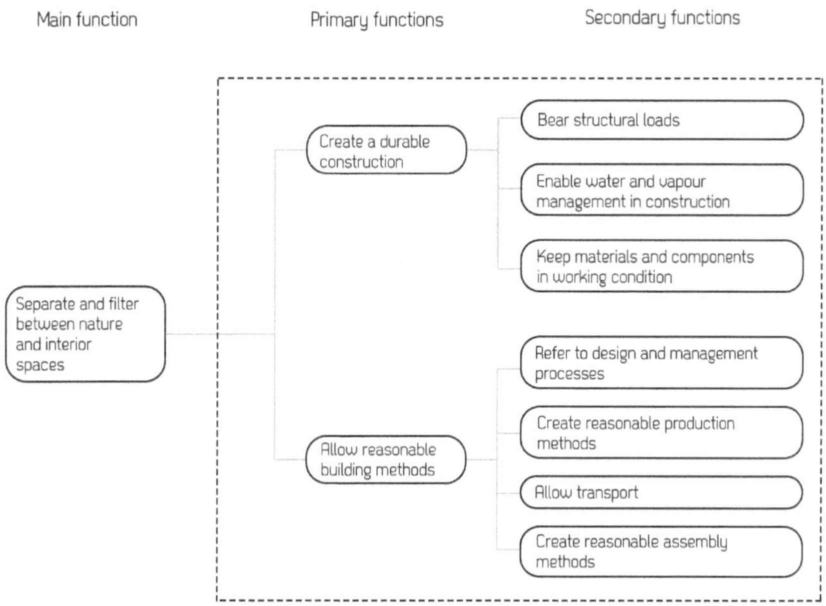

Figure 2.30 The façade function tree (Klein 2013).

Curtain walls are one of the most prominent façade systems. The architects and engineers have become familiar with how to utilize these systems in their designs. And yet, continuous innovation in form-finding and structural techniques as well as the continuous passion of the architects for more individual and unique designs, has increased the tendency towards individualized solutions. The limitations that are introduced to any new product are mainly defined through process planning.

Process planning is the transition from design to physical product, and it defines the complexity and the limitation of the final product. Mp. Groover defines process planning this way, "Process planning. As our definition suggests, this is the principal activity of manufacturing engineering. Process planning includes (a) deciding what process and methods should be used and in what sequence, (b) determining tooling requirements, (c) selecting production equipment and systems, and (d) estimating costs of production for the selected processes, tooling, and equipment."(Groover 2010).

Façade constructions have developed steadily over the past decades. These developments were forced into the building industry after the energy crisis. As a result, the standard façade systems were able to achieve excellent performance after being improved through a long period of testing and certifying. Still, these systems are limited, based on industrialization aspects. The limitations of any façade system are divided into three: structural limitation-when the design present severe loading conditions that the systems in question cannot handle-, production limitation-when the production of the element itself cannot be standardized-, and cost/time limitation-when the product has weak performance-cost ratio-.

Limitations on the structural level come from two factors: the geometrical conditions of the whole structure and the structural integrity of the building components.

The geometrical conditions that the connecting node must accommodate are based on the free-form or complex envelope that has to be realized. These geometrical conditions are: the joinery form between the node and the profiles, the size of the profiles and their orientation, and the variable number of the connecting members in some cases. The structural integrity has to be achieved, which affects the material choice and the size of the node. These factors influence the decision regarding which specific method will be used, such as machining, welding/fabrication, casting, and so forth.

The limitations on the production level are defined by means of the chosen method of manufacturing and its sequence. The initial geometry of the connecting element is manufactured through the primary process. Each secondary process then increases the level of complexity required to manufacture the part. Finally, finishing operations may be needed to treat the parts' outer surface. These shaping processes may be subject to sub-operations to enhance properties, which are additional steps that extend the whole process.

All aspects mentioned above are the main factors of the cost estimation process. The design process, represented in the working hours, is needed to finalize the requested documents that describe the part thoroughly. Quality requirements have to be fulfilled, such as tolerances, repeatability, and other quality measures. Production volume influences the cost estimation dramatically, in terms of production rate and selection of processes. Material utilization detail is desirable, and it affects the sequence of the process to achieve efficient material usage and reduce costs. Finally, the number of separate steps has to be optimized to save time and reduce material handling (Groover 2010).

2.5 Objectives

The challenges presented here serve to prove that a new strategy is needed with regards to complex building envelopes that exceed the standard systems' limitations. This strategy will define the future of curtain walls and will open new possibilities. The objective of this dissertation is the utilization of a file-to-factory concept to design and engineer complex building envelopes. This is done by the integration of additive manufacturing and parametric design into the planning process.

Parametric design will improve the manufacturability of façade components and will help organize the data set for a project. Additive manufacturing will open an extra range of possibilities to standard façade systems, and it will shorten the whole process. Additive manufacturing can be seen as a complement to the standard façade systems.

To implement any new building component in the building/façade industry, a series of testing and certifying must be considered. Additive manufacturing has no certification that allows it to be used in the façade industry. This means that the additively manufactured elements have to be tested as individualized solutions. These tests will be investigated.

The process will be presented as a design and engineering concept that could be adapted to many geometrical conditions. This is limited to the capabilities of the chosen standard system.

3 Methods of manufacturing | AM of Metals

Additive manufacturing of metal alloys is a fabrication process, where a three-dimensional object is built by adding layers on top of each other through laser scanning of a powder bed. This 3D printing family group has different categories, such as SLM (Selective Laser Melting), SLS (Selective Laser Sintering), and EBM (Electron Beam Melting). These methods differ in the fundamentals of the bonding process between the particles, and the melting degree.

In this dissertation, the focus is given to the SLM method. This focus is driven by the melting degree that SLM grants: 100% melt of the powder particles (Full Melting) as well as the usage of pure metal powders with no addition of any bonding agents (Kruth, Mercelis, Van Vaerenbergh, Froyen, & Rombouts, 2005).

3.1 SLM (Selective Laser Melting)

Selective laser melting is premature manufacturing method in comparison to more conventional manufacturing methods like casting, molding, joining, coating, and machining. These methods belong to Subtractive Manufacturing family, where 3D objects are constructed by cutting materials away from a solid material block. SLM belongs to Additive Manufacturing family, where 3D objects are transformed into layers and constructed by laying them one above the other till the targeted object is formed.

The functionality of SLM is described through three stages (Figure 3.1): a) a thin layer of metal powder is spread on a base platform or substrate, through a silicon lip, metal blade or metal roller. b) A laser beam is conducted through a scanning mirror on specific areas of the powder bed. These areas present the shape of the layer. This shape will be fully melted through the focused laser. c) After the completion of the current layer, the build plate will be moved down with the defined layer thickness to repeat the first step (Löber, 2014).

A. Mohsen, *Design to Manufacture of Complex Building Envelopes*, Mechanik, Werkstoffe und Konstruktion im Bauwesen 56,
https://doi.org/10.1007/978-3-658-30204-7_3

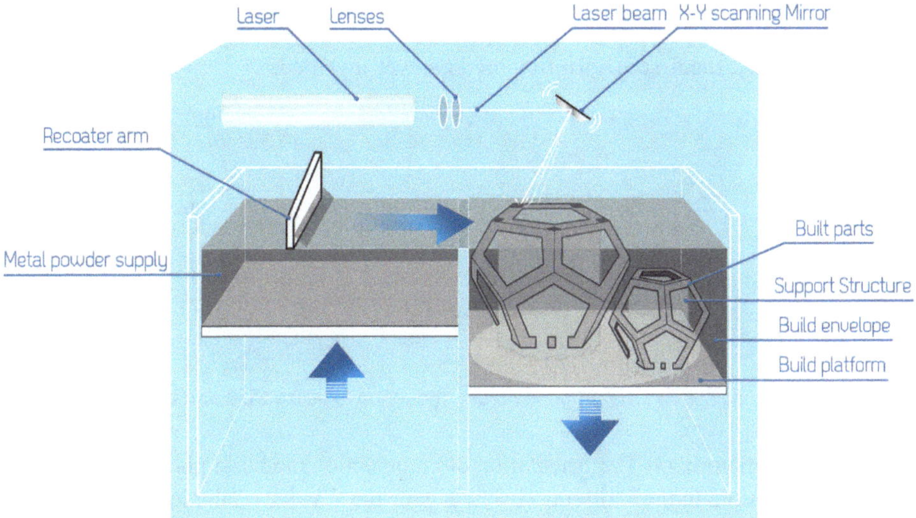

Figure 3.1 Schematic shows the functionality of the SLM process. Additively. (2018). Laser Melting. Retrieved from https://www.additively.com/en/learn-about/laser-melting

3.1.1 Sequence of utilizing SLM for façade parts

To implement the SLM process in the building industry, a sequence of testing and simulations is investigated to grant the fulfillment of the manufactured parts for the building regulations.

The sequence of the process is defined through a series of steps (Figure 3.2). Each step presents certain processes that varied between real testing and digital correction for the requested parts. The first step is selecting both the SLM machine and the metal powder. SLM machines are controlled through certain parameters. Some of these parameters can be controlled, and others are direct results that come out of the physical interaction of the process. The metal powder affects the characteristics of the printed part. Therefore, its properties must be considered.

To design a façade element, the mechanical properties of the part have to be known to predict the part's behavior. As a result, tensile specimens are tested using DIN 5012, the standard that specifies all the requirements for tensile tests on metallic materials (Norm, 2016). Then validation models are designed and tested to tweak the results of the tensile testing and to improve the simulation process. Once precise mechanical properties are defined, an object can be designed and engineered via any CAD software to prepare for the printing stage. ".stl" format delivers the needed data that the 3D printing machine needs. .stl is a shortcut for "Standard Tessellation Language" or "Standard Triangle language," which is the standard file format for all 3D printers worldwide as it was first generated for Stereolithography, the first 3D printing method ever invented.

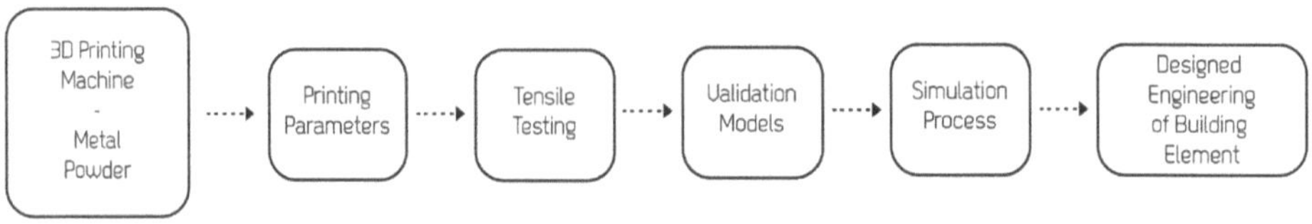

Figure 3.2 Schematic shows the proposed tests' sequence to utilize SLM parts.

As a rule of thumb: (1) The orientation of an object on the build plate has to have the least height possible in the Z direction. The bigger the Z value is; the longer the printing process is, the more material is used, and the more complex the post-processing is. (2) Any side of the object that tilts more than 45° to the build plate will need support structures, which are not volumetric parts, but rather melting paths. That grant the success of the print.

At this point, the process's printing parameters such as layer thickness, laser speed, laser power, hatch spacing, and exposure parameters, are defined. Finally, the object is separated from the building plate and is ready for post-processing, which includes removal of the support structure, sanding, and coating.

3.2 Characteristics of the SLM process

3.2.1 Parameters of the SLM process

There is an overview of more than 50 factors that may influence the SLM process, investigated by Maarten Van Elsen in his dissertation "Complexity of selective laser melting." These parameters are Environment, Powder composition, Powder granulometry and deposition, Powder bed properties, Laser, Processing parameters, and constants. (Van Elsen, 2007)

- **Environment**
- Inert gas: Each inert gas reacts differently to the melt. Non-inert particles like oxygen can react with the molten metal.

- Molecular mass: The molecular mass influences the heat balance.
- Viscosity: The surrounding gas viscosity may have a minimal impact on the free surface activity of the melt pool.
- Thermal conductivity: The conductivity of the surrounding gas influences the heat balance.
- Heat exchange coefficient of convection: Convection is present but negligible.
- Thermal capacity: The heat capacity of the surrounding gas influences the heat balance.
- Pressure: Chamber pressure is negligible.
- Oxygen level % O2: Oxygen level plays a huge rule in the SLM process, as in the presence of oxygen; metal will oxidize, which affects the object properties. (Hauser, Childs, Dalgarno, & Eane, 1999)
- Ambient temperature: The ambient temperature affects the printed object, as the preheating of the powder or the building plate influence the object properties.
- Surface free energy: The shape of the melt pool is affected by the surface free energy between the liquid and the surrounding gas.

• **Powder composition: Material**

- Bulk density: The density of the bulk material determines the maximum final density of the component and the powder material.
- Thermal conductivity: The conductivity of the bulk material determines the conductivity of the powder material.
- Thermal capacity: Thermal capacities of the material influence the heat balance.
- Latent heat of fusion: The amount of energy needed to be absorbed by the material to change its phase, which will be freed later during solidification.
- Melting temperature: The effect of melting temperature is integrated into the total la-tent heat of fusion because alloys do not have a single melting point.
- Boiling temperature: The boiling temperature should never be reached to avoid material evaporation.
- Dynamic viscosity: The dynamic viscosity has a significant influence on the melt pool dynamics like balling, wetting and Marangoni convection.
- Thermal expansion coefficient: The thermal expansion coefficient is important for the stresses introduced during solidification and cool down (Nickel, Barnett, Link, & Prinz, 1999).
- Surface free energy: The surface free energy is a function of the two media in contact. Accordingly, it quantifies the disruption of chemical bonds when a surface is trans-formed, and it determines the wetting and balling behavior of the melt pool.
- Vapor pressure: It is important when evaporating occurs, which is not desired during SLM.
- Reaction energy: It is not clear how this factor influences the SLM process.

- Absorptivity: It is the measure of the amount of laser energy that is absorbed by the material compared to the amount of energy that is transmitted and reflected.
- Diffusion coefficient: Unlike the sintering process, the diffusion coefficient has no significant role in SLM.
- Solubility: The solubility of the solid material in liquid material might influence the process of SLM, especially when the powder particles are attracted to the melting pool.
- Melt enthalpy: The amount of energy needed to melt the powder and reach to the complete melt state.
- Pollution: When the powder is reused it is subject to dust pollution. Therefore, pollution could affect the SLM process, and it should be kept under control.

- **Powder granulometry and deposition**
- Morphology: Powder morphology refers to the shape of the individual particles. Spherical particles deliver the best flowability and have a higher random packing density.
- Surface roughness: The roughness or surface quality of a single grain of powder mate-rial influences the flowability and affects the absorptivity of the powder.
- Particle size distribution: The particle size distribution is the relative frequency of particles of a given size and a different probability of such distribution – Gaussian / bimodal – could be used to achieve the best packing density.
  The maximum particle size determines the minimum layer thickness. Small particles can fill the void between the bigger particles. Thus, higher packing density could be reached, but particles that are too small have negative properties as well. Nanoparticles can penetrate through the skin and when inhaled, can diffuse through the components of the equipment, and they can be easily displaced from the melt pool.
- Deposition system: The powder transfer from its container to the building platform is done through a deposition system. Each system may have different influence over the powder deposition.

- **Powder bed properties**
- Density: Comparing the powder bed density to the bulk density indicates the compaction needed to obtain fully dense components.
- Conductivity: The conductivity of the deposited powder influences the heat balance.
- Absorptivity: The wavelength of the laser and the state of the powder bed define the absorptivity of a powder. This refers to the amount of absorbed laser energy by the powder bed, as the rest is reflected in the processing chamber.
- Emissivity: The hotter the melt pool, the more energy is lost by radiation, which influences the heat balance.
- Diffusivity: It is a frequently used property in heat transfer and could determine

the penetration depth of the heat.

- **Laser**
- Mode: Laser systems can be operated in two different modes: continuous and pulsed waves.
- Wavelength: The wavelength of the laser determines the absorptivity.
- Intensity profile: The intensity profile is an important factor for the SLM process, as it is responsible for determining how much heat is added in a specific place. However, it cannot be freely chosen.
  The next properties are a summary of the factors for the intensity profile.
- Average power: The total amount of added heat.
- Peak power: The maximum power for pulsed laser.
- Beam quality: Predicts how well the beam can be focused, accordingly determines the theoretical minimal spot size of the laser system.
- Frequency: The frequency of the laser pulses per time cycle and how peak power is determined.
- Pulse width: With the frequency, the peak power is determined.
- Spot size: For an elliptical laser beam, spot size is defined by d_x and d_y. Only circular spots will be isotropic in different directions.
- Polarity: The effect is unknown.

- **Processing parameters**
  The energy source is very fine. Therefore, the layer formation is done through a scanning pattern of lines. The settings of this pattern influence heat balance, building speed, and accuracy.
- Scan Strategy: The method to fill the required cross sections with scan tracks. The cross section is subdivided into quadrants and sectors; then they are scanned in certain order to be filled with certain patterns, such as zig-zag, hatching, or spiral. The scan strategy affects temperature disruption.
- Scan spacing: The distance between the scan tracks; the smaller it is, the slower the process.
- Scan speed: The faster the scanning speed, the lower the building time, it also influences the heat balance.
- Energy density: Depending on the parameter settings, more heat is added to the powder bed.
- Layer thickness: The thickness of the deposited layer determines how much powder will melt during a single scan, and it is bigger than the processed layer thickness according to the shrinkage that occurs because of the low density of the powder.
- 3D environment: The geometry of the printed part influences the process.

- **Constants**
  - Gravitational acceleration: $g \approx 9.81$ m/s^2
  - Constant of Planck: $h \approx 6.626 \cdot 10^{-34}$ J·s.
  - Constant of Boltzmann: $k_B \approx 1.381 \cdot 10^{-23}$ J/K.
  - Universal gas constant: $R_G \approx 8.314$ JK^{-1} mol^{-1}.
  - Speed of light: $c \approx 3 \cdot 10^8$ m/s.
  - Radiation constant of Stefan-Boltzmann.

All these parameters may or may not influence the 3D printed object. The significant parameter from the environment is the inert gas as it the first medium that interacts with the molten metal. From the powder composition; the melting temperature helps to define the machine capacity. Powder morphology influences the molten pools (see 3.2.3). The main processing parameters are laser power, laser speed, hatch distance, and layer thickness. Each manufacturer usually has his own set of processing parameters that work best for his material.

In building/façade industry, any parameter that affects the mechanical properties of the element is of significant interest. To better evaluate the weight of the parameters, the interactions through the process have to be clarified and investigated.

3.2.2 The Interactions of the SLM process

The absorption rate of the laser rays is a critical physical aspect of SLM. It is defined by the polarization and the wavelength of the laser rays, as well as the applied metal powder and its properties. The absorption rate of laser rays on a powder layer is significantly different from the absorption rate of laser rays on a homogeneous solid body. The laser rays allow an additional scattering effect to occur on the powder layer. The scatter, and the absorption of electromagnetic rays on spherical particles are related to the ratio between the particles' size and the rays' wavelength (Löber, 2014).

Mie-Parameter Q is used as a representative value for this ratio. (Equation 3.1) determines the Mie-Parameter Q (Mie, 1908):

$$Q = \frac{\pi dk}{\lambda} \tag{3.1}$$

where:

d_k = particle diameter

λ = wave length

Dependent on q value, three zones could be distinguished:

- q << 1 Rayleigh scattering
- q ≈ Mie scattering
- q >> 1 Ray optics – Geometrical optics –

Based on the reflection angle of the laser rays on the powder layer; the laser rays can either entirely reflect away or pass through to a deeper layer (Figure 3.3). The sum of the reflected, transmitted, and absorbed laser rays has to be equal one. Accordingly, the following balance equation is generated:

$$A = 1 - (R+T) \tag{3.2}$$

where:

A = absorptivity
R = reflectivity
T = transmissivity

The absorptivity of metal powders lies within a range of 45-75%. Aluminum lies in a lower range, while Titanium lies in an upper range. The transmissivity usually lies below 1 % (Xie, Kar, Rothenflue, & Latham, 1997).

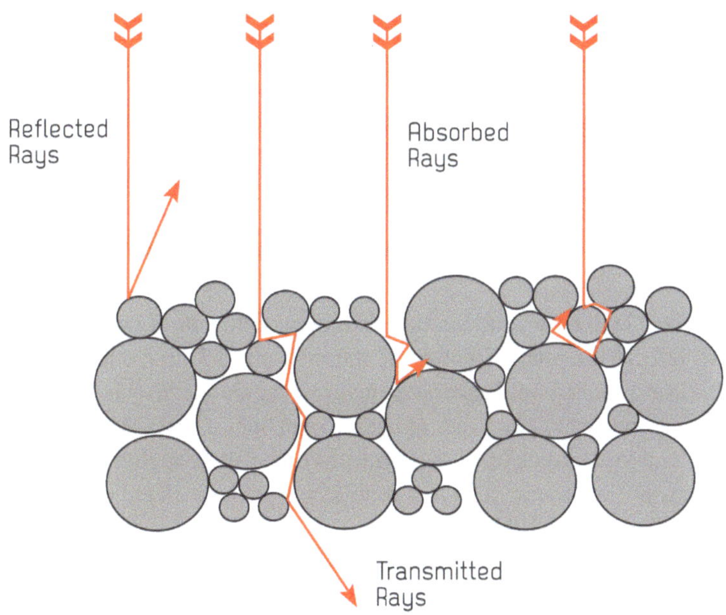

Figure 3.3 Schematic showing the interactions between the laser rays and the metal particles during the SLM process (Löber, 2014).

The cooling of the melt by laser welding is defined as:

$$\frac{dT}{dt} = -2\pi\lambda\left(\frac{v}{P}\right)\Delta T^2 \tag{3.3}$$

where:

λ = Thermal conductivity
v = Laser speed
P = Laser power
ΔT = Temperature difference between the substrate and the melting pool (Wilhelm, 1999)

This equation clarifies the effect of increasing the substrate temperature. This increase allows for the reduction of the cooling speed. This reduction allows for the adjustment of the morphology, distribution, and the size of the formed phases. Furthermore, there are other parameters out of kinetics that define the solidification art and the effect on the microstructure. The velocity of Solidus-Liquidus-boundaries belongs to the effects on the microstructure; also known as growth rate, or the temperature gradient between the molten metal and the Solidus-Liquidus-boundaries. For pure metals, the solidification under equilibrium condition is planar. Only when extreme supercooling is forced can the solidification art be controlled. (Löber, 2014)

3.2.3 Solidification of the melt during the SLM process

SLM is characterized by the short interaction time between the laser rays and the powder particles, which affects the microstructure of the melt pool. The overlapping of multi-track and multi-layer molten pools generates the layers/components. The microstructure of these layers/components has two types of molten pool boundaries (MPBs). These types are "layer-layer" and "track-track" MPBs (Figure 3.4). The additive rhythm and the solidification conditions of the process lead to a unique morphological and crystallo-graphic microstructure (Shifeng et al., 2014).

The noticeable mechanical properties of SLM parts are high strength, anisotropy, and low ductility. For example, the yield and ultimate tensile strengths of Inconel 718 components slightly exceed those of annealed/wrought components, whereas the elongation is approximately the same (Amato et al., 2012). The yield stresses of 316L stainless steel,

CoCr alloy, and Ni625 increased by almost 50%, and the elongation by almost 20-30%, in comparison to the forgings (Yadroitsev, Yadroitsava, & Smurov, 2011). 17-4 PH stainless steel presented an exceptional increase in yield stress when hardened from 600 MPa to 1300 MPa, and a total elongation of fracture of 28% (Facchini et al., 2010). The tensile strength of Ti-6Al-4V was 50% greater than forgings (Murr et al., 2009).

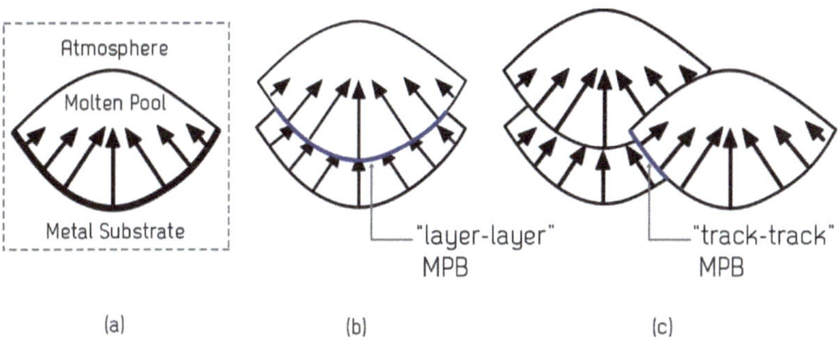

Figure 3.4 Schematic of the crystallization of SLM melting pools (Shifeng et al., 2014).

"Layer-layer" MPBs are the ones generated from an overlapping alongside the Z-axis layer by layer. "Track-track" MPBs are the ones generated from an overlapping alongside the X-Y plane (Figure 3.5). The density of SLM parts can be controlled by the adjustment of the overlapping of the molten pools alongside the Z-axis and the horizontal X-Y plane. The solidification conditions define the grain orientation. The powder grains form the liquid pools when melted by the laser. The molten pools have approximately arcuate cross-sections. The upper half of the molten pool is exposed to the atmosphere, where the heat flux is transferred in a radiation mode. The lower half contacts the solidified metal substrate, where the heat flux is transferred in a conduction mode. Non-Homogeneous nucleation on the solid phase boundary at the bottom is easily formed; this allows the fast transfer of the heat flux in a perpendicular direction to the solid-liquid interface. This allows the grains to grow alongside the opposite direction as shown by the arrows (Figure 3.4). The solidification rate is rapid, and is related to the laser scanning speed (Shifeng et al., 2014).

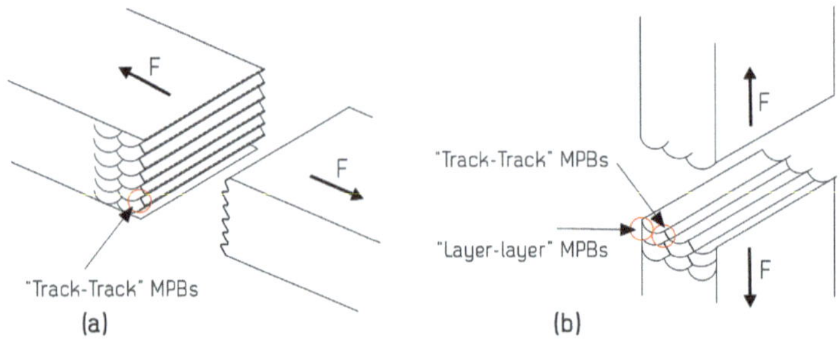

Figure 3.5 Schematic of the relation of SLM melting pools (Shifeng et al., 2014).

3.3 Characteristic of Metal Powders

3.3.1 Powder production techniques

There are several methods concerning metal powder production, which can be classified as follows:

1. Chemical methods
2. Physical methods
3. Mechanical methods

Chemical methods

Chemical methods can be further classified into chemical reduction and decomposition.

- **Chemical reduction**

Chemical reduction is one of the essential generation processes of metal powders. This process involves a chemical compound such as an oxide or a halide. The chemical reduction process can be carried out in three different states.

a) From the solid state, where the reduction is done of iron oxide with carbon, or of tungsten oxide with hydrogen (Figure 3.6). The typical example of this production method is the Höganäs process, where the iron ore is heated in the presence of coke and limestone. The result is the sponge iron, which is mechanically removed and ground. The powder is then magnetically separated from impurities (Schatt, Wieters, & Kieback, 2007).

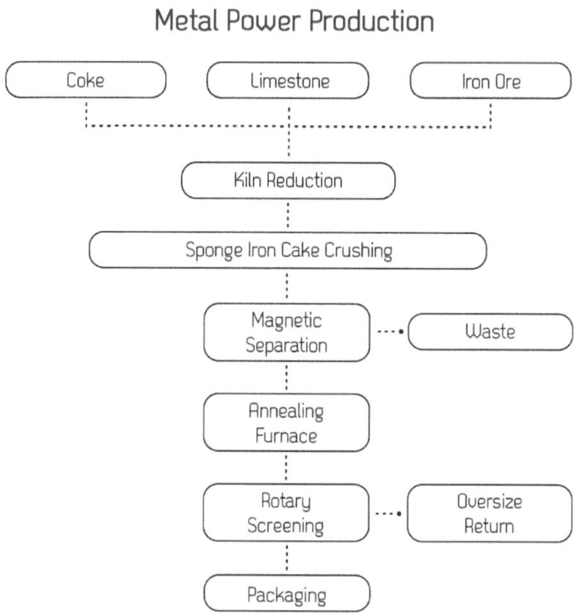

Figure 3.6 Schematic of the process of sponge iron powder production from iron ore (Schatt et al., 2007).

b) From the gaseous state, where the reduction is made of titanium tetrachloride vapor with molten magnesium. The process is called Kroll process.
c) From the aqueous solution, where the precipitation is made of cement copper from copper sulfate solution with iron or the reduction of an ammoniacal nickel salt solution under pressure with hydrogen (Upadhyaya, 1997).

- **Chemical decomposition**

The powder production category has two prevalent methods. The first method is the decomposition of metal hydrides, where the refractory metals, such as Ti, Zr or Hf, are hydrided by heating in the form of a sponge, chip, or turnings. These hydrides are brittle and can be ball-milled into powder. The second method is the decomposition of metal carbonyls, where the carbonyls that are liquids at normal temperature are decomposed by heating the vapor at atmospheric pressure (Upadhyaya, 1997).

Physical methods

Electrolytic method and atomization are the conventional physical methods for powder manufacturing.

- **Electrolytic method**

This approach is widely used for the preparation of copper, beryllium, iron, and nickel powders. Electrodeposition allows the adjustment of the chemical and physical conditions (Figure 3.7). This adjustment permits the metal deposition on the cathode of the cell. This metal deposition is either as a light cake or in a flake form. Both forms are then crushed into powder. The generated powders/deposits characterize with high purity, reactivity, brittle, and excellent properties for conventional powder metallurgy processing. The formed powders have a dendritic shape; however, this can be changed through post-processing. The electrolytic iron powder has a higher green density than reduced or atomized powder (Schatt et al., 2007; Upadhyaya, 1997).

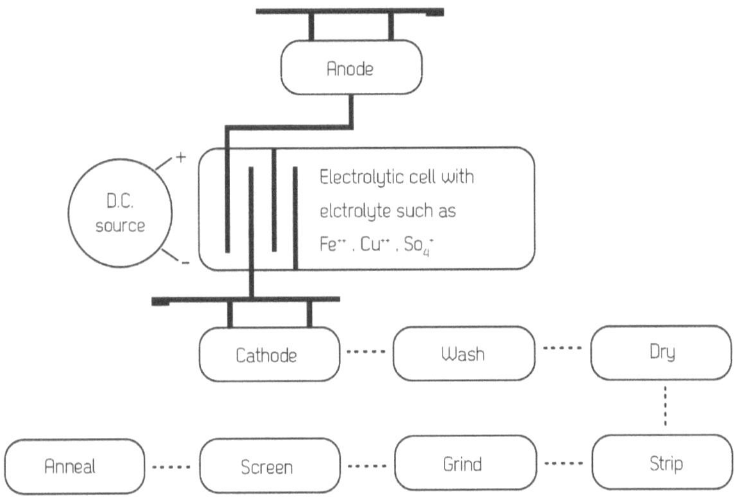

Figure 3.7 Schematic of the electrolytic process of metal powder making (Schatt et al., 2007).

- **Atomization**

Atomization is based on the theory that any material can be made into powder by the disintegration of its melt. Atomization has three categories. These categories are gas, water, and centrifugal (Lawley, 1978). Atomization is widely used because of its ability to produce high purity powders. The production is done through forcing the melt through an orifice, then inducing turbulence in the melt through the impinging gas or liquid.

The gas atomization is done through the use of air or inert gases, such as nitrogen and argon. The mixing process is either: 1) external mixing, where the contact between the melt and the atomizing medium takes place outside the nozzle, which suits the atomization of metals, or 2) internal mixing, where the contact between the melt and the atomizing medium takes place inside the nozzle, which suits the atomization of materials that are liquids at room temperature. This atomization method often delivers spherical powders; yet it is expensive and not energy efficient (Schatt et al., 2007; Upadhyaya, 1997).

Water atomization is done through a high-pressure water stream that is forced through a nozzle. As a result, the water forms disperse droplets that impact the metal stream. This atomization method delivers spattered powders with oxidized surfaces, and it is not energy efficient (Schatt et al., 2007; Upadhyaya, 1997).

Centrifugal atomization is done through the ejection of the melt from a rapidly spinning container, plate, or disc. The action of the centrifugal force generates the melt. Two examples of the process are the rotating electrode process (REP), and the plasma rotating electrode process (PREP). In the two processes, the material is the rod electrode that is melted at one end through an electric or a plasma arc. This atomization method is mainly developed for high purity, low oxygen titanium alloys and superalloys (Roberts, 1984; Schatt et al., 2007; Upadhyaya, 1997).

Mechanical methods

Mechanical methods rely on interactions, such as crushing, grinding, cutting, and compression. The mechanical method is used when materials are relatively easy to fracture, or when they are needed in a flake form. The standard method is a ball mill that consists of a rotating drum. The rotating drum mill must have wear-resistant balls in that are strong in comparison to the material that they press against. An optimum speed has to be achieved to grant a continuous flow of the process (Upadhyaya, 1997).

3.3.2 The properties of Metal Powders

Each metallurgical powder process affects the characterization of the produced metal powder. The characteristics of produced metal powders are represented in the subsequent particle chemistry and structure. The particles' properties influence the behavior of the powder during sintering, melting, and compaction. Chemical composition of the particles is one of the properties that significantly influence processing and the final product. Chemical composition is defined by the level of impurity. These impurities have to be defined, whether they are present in an elemental form or a chemical compound. For example, silica is present in reduced iron powder as a form of impurity. This kind of impurity influences the hardness of the particles, the degree of chemical reactivity, the microstructure of the crystalline powder, and the grain size. Accordingly, these influence the properties of the final product (Upadhyaya, 1997).

The metal powder's shape is defined by its particle's dimensionality and contour surface. The particles of most powders are three dimensional and considered equiaxed (Figure 3.8). Porous particles may be present, which makes shape characterization very difficult. In practice, the separation and classification of powders are mainly done by sieving (Schatt et al., 2007; Upadhyaya, 1997).

Figure 3.8 Schematic of particle shapes' characterization.

Particle surface topography is defined by the roughness degree of the particles. This roughness degree defines the particles' surface quality. Surface topography is examined by scanning electron microscope; it influences the frictional forces between the particles, the bulk movement of the particles, and the actual contact particle-to-particle during sintering (Upadhyaya, 1997).

The specific surface area of metal powders is a significant characteristic that affects sinterability. The absorption method from S. Brunauer, P.H. Emmett, and E. Teller (BET-Method) is considered the primary measuring method to define the specific surface area of metal powders (Kaye, 2008; Norm, 1993; Schatt et al., 2007; Schulze, 1994).

Apparent and tap density refer to a mass of unit volume that is expressed in g/cm^{-3}. Different measuring methods of the density of metal powders are discussed in DIN ISO 3923, parts 1, 2 &3. The density of metal powders determines the selection of equipment that is suitable for transportation, the treatment, and the behavior of powders during sintering (Schatt et al., 2007; Upadhyaya, 1997).

The flow rate of metal powders is defined by the time needed for 50 g of powder to flow through a small standardized orifice called Hall flowmeter, as described in DIN ISO 4490. Powders with small size particles have a bigger specific area which leads to a higher apparent density, which lowers the flow time (Schatt et al., 2007; Upadhyaya, 1997).

Green strength is the mechanical strength of the green compact, which is the unsintered powder compact. It is crucial, as it defines how the green compact maintains its size and shape while prepared to sinter (Upadhyaya, 1997).

Each metal powder treatment has its influence on the sintering/melting process, and affects the final product in a certain way. These treatments are, e.g., annealing, mixing, size reduction, granulation, and metal powder coating.

The spherical particle shape is most suitable for 3D printing of metals because of the homogenous molten pools (see 3.2.3).

3.4 Characteristic of SLM parts

3.4.1 Testing and Analysis Methods

Tensile testing

Tensile testing is one of the bases for material science and engineering, where a controlled tensile force is applied to a sample until failure. It is used to define some of the mechanical properties of the material, such as Young's modulus and breaking strength. Tensile testing is usually done at room temperature, which is approximately 23°C in accordance to (ISO 6892, 2019).

Tensile testing is deemed one of the most important and recognized testing methods, when it comes to defining the tensile strength of materials. The main scientific reasons are: a) it provides material parameters that are used for calculation processes, b) the appliance of law of similarity, which allows the usage of the defined values for the calculations to any other construction, c) through years of testing, the values allow conversion possibilities to generate other material properties (Riebensahm, 2013).

In a tensile test, the specimen is designed with respect to the user testing machine. This means the integration of individual parts to the specimen to work at its best with combination to the clamping tools, such as screwed grips, parallel jaw faces, wedges, or shouldered holders. The following are essential: to ensure that the forces act as axially as possible on the clamped specimen, to keep the bending effect to a minimum, and to set the force-measuring system to zero after assembly (ISO, 2017).

Tensile specimens' cross-section varies between circular, quadratic, and rectangular. The used cross-section is regulated in DIN 50125 (Norm, 2016). The norm facilitates the choice of the form and the method of manufacturing.

Testing rates are categorized into two types (Figure 3.9): Strain rate (method A), and Stress rate (method B). Method A is subcategorized into two types: method A1 (closed loop) that involves the control of the strain rate itself, and method A2 (open loop) that involves the control of the estimated strain rate over the parallel length. Method B involves the control of the stress rate, then the measure of the accruing strain (ISO, 2017).

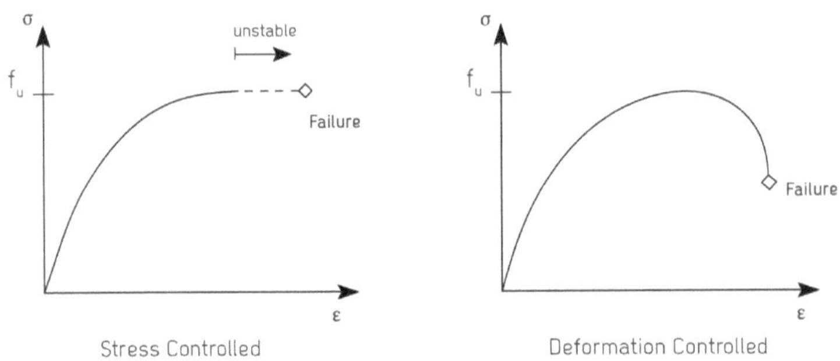

Figure 3.9 Tensile testing methods (Method A, right – Method B, left).

Stress conditions

In building components and structural elements—depending on the load case and the component/element geometry—occur uniaxial, biaxial, shear or triaxial stress states. The uniaxial stress state is defined as only one stress component is acting on an infinitesimal small volume element. All other stress components are equal to zero. Hence, a uniaxial stress state is apparent. If two stress components are acting on an infinitesimal small volume element, a so-called biaxial stress condition is apparent, since the third stress component is equal to zero. Considering the case that three stress components are acting on the volume element, a triaxial stress state is present. To obtain shear state in the volume element, the stress components on the principal diagonl of the stress tensor are equal to one, whereas shear stresses are acting on the secondary diagonal of the stress tensor.

Digital Image Correlation (DIC)

In engineering and building science, it is important to determine experimentally the force-displacement and stress-strain behavior of a material or structural component. In order to characterize a material mechanically, it is therefore essential to determine the deformation behavior as well as the local behavior resulting from the measurement of strain.

DIC is a non-contact optical measuring technique that is a simple to use and cost-effective in-situ measurement. It works through digital photography tracking of the component, at different stages of deformation (McCormick & Lord, 2010).

In traditional strain measurements technique, discrete data is given by averaging the real strains between the two ends of the used strain gauges. Traditional strain measurements are known to be reliable and robust, although they are affected by the environment, such as the preparation of the surface, machine condition, and wiring (Dally & Riley, 1965).

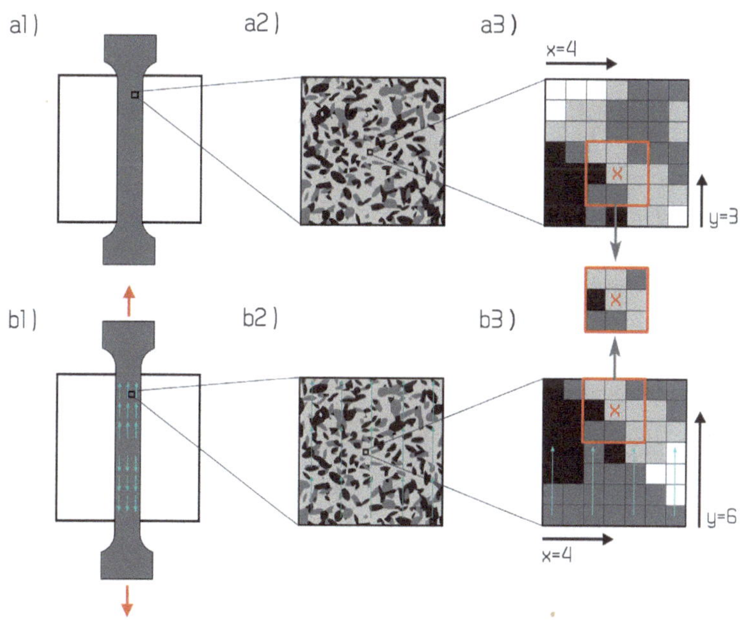

Figure 3.10 Principles on digital image correlation.

On the contrary, DIC permits the three-dimensional measurement of local elongation and displacement of complex geometries. This measurement is done by capturing the full field strain data with good resolution images of the specimen's surface (Peters & Ranson, 1982).

The principle of DIC is to pattern a sample stochastically with color and then to follow the individual positions of the so-called speckles, which allows one to derive the deformation gradient and thus the strains (Figure 3.10). This outline defines the first module that is a grid based on the strain level required for the analyses, and the second module that is the core of the correlation tool. The core of the correlation then translates into a set of pixels a grey scale level, to correlate a part of the source image with the target image. A reference image of the unstrained specimen is taken, which shows the required structure needed for the analysis, then the software defines the pixel-subsets in grey scale levels. The process is repeated on a constant speed over time, and finally, the correlation algorithm defines the displacement of all points. The software can cover all the shown surface and provide a detailed analysis of the elongation distribution over the specimen.

Finite Elements Method (FEM) and ANSYS Simulation

The basics of the Finite Elements method was introduced by Turner er al. in 1965. These basics helped to introduce a powerful computational technique for approximate solutions to different "real-world" engineering problems. The technical simulations permit the visual analysis to test specimens and building components. These simulations save physical prototyping and testing. The simulations try to mimic reality, which both elaborates the development cycle and provides more design freedom and safety (Madenci & Guven, 2015).

The different steps in the utilization of FEM are modeling of the problem, simplifying the model, formatting domains and subdomains, and developing a solution (Figure 3.11). The first step shows the modeling of the problem, which represents the actual model of the building component. The three-dimensional model is simplified to its two-dimensional plane. The domain is decomposed into a finite number of subdomains (FE-Net). For each element, an approximate systematic solution is constructed. The element matrices for each subdomain are assembled to obtain the global matrix of the entire domain. Each element allocates the required thickness in order to consider the third dimension. Then the boundary conditions are defined. The stiffness matrix of each element is established, then assembled to the global stiffness matrix. The global stiffness matrix (**K**) is a relation between force vector (F) and displacement vector (u), (Equation 3.4). (Madenci & Guven, 2015; Steinke, 2015).

$$\vec{F} = \mathbf{K}\,\vec{u} \tag{3.4}$$

where:

$\vec{F}$ = Force
$\mathbf{K}$ = Global stiffness matrix
$\vec{u}$ = Displacement

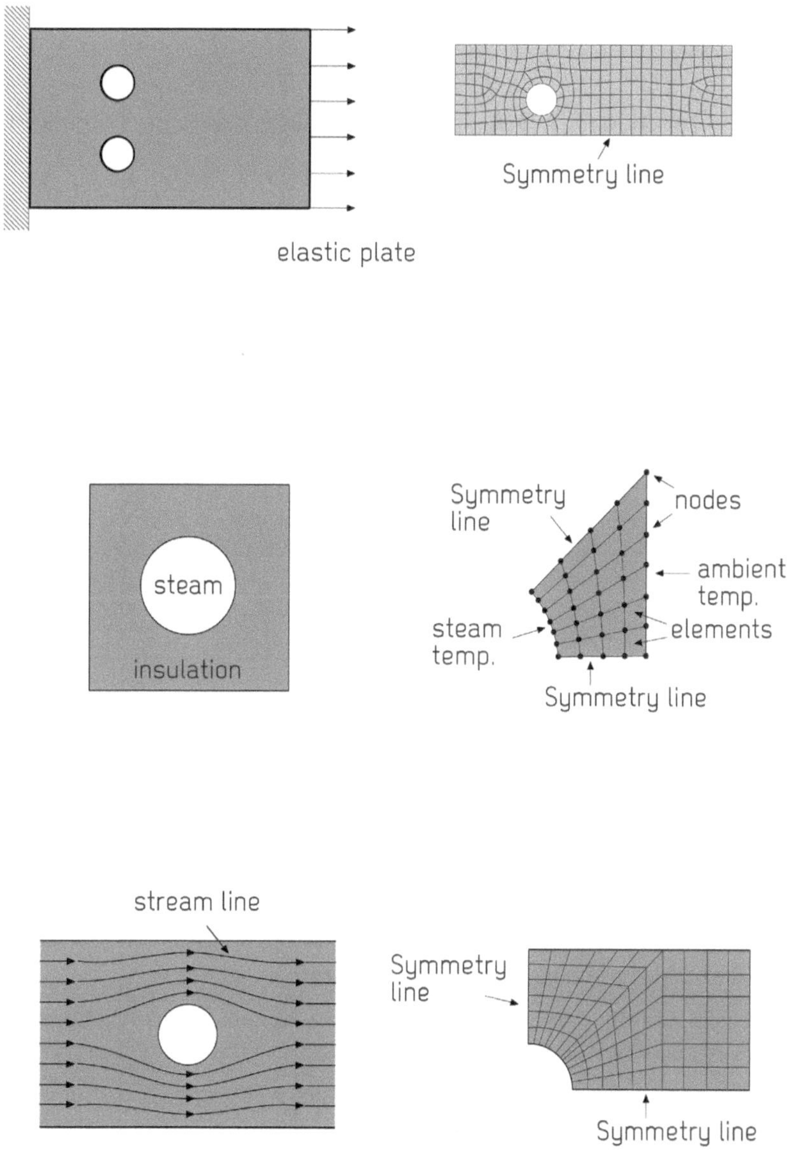

Figure 3.11 Schematic shows FEM interpretation (Madenci & Guven, 2015).

The process can be subdivided into three stages: a) Pre-processing, b) Solution, and c) Post-processing. In the pre-processing, the geometry, material properties, grid type, and generation of calculation grids are defined. Then the boundary conditions are defined, and the model equations are executed in the solution, and finally, the visualization/evaluation of the results is shown at the post-processing (Westermann, 2010).

3.4.2 Testing and Analysis of Mechanical Properties

As shown in section 3.1.1, the general process starts with choosing the 3D printing machine. In this research, two different machines have been used to manufacture the test specimen. EOS M 290 400W was available at the Institut für Produktionsmanagment, Technologie und Werkzeugmaschinen at the Technical University Darmstadt, and was used to manufacture the first test cycle – made of Stainless-steel 316L (1.4404) (see Appendix C). The powder is gas atomized, and it is the commercial version from the same manufacturer. The machine combines a 400 W YB fiber laser with 84 µm spot size. It operates using Argon as an inert gas, and has a building volume of 250 mm x 250 mm x 325 mm (see Appendix A).

The applied print parameters of the printing process for the Stainless-steel (316 L) testing specimens are defined (Table 3.1).

Table 3.1 Printing Parameters – Stainless-steel (316L).

Laser Power	195 W
Laser Speed	1080 mm/s
Hatch Distance	0,09 mm
Layer thickness	20 µm

Trumpf machine (see Appendix H) was used to manufacture the second test cycle that was made of Aluminum powder "AlSi10Mg" (see Appendix G).

The applied print parameters of the printing process for the Aluminum (AlSi10Mg) testing specimens are defined (Table 3.2).

Table 3.2 Printing Parameters – Aluminum (AlSi20Mg).

Laser Power	420 W
Laser Speed	1300 mm/s
Hatch Distance	0,2 mm
Layer thickness	60 µm

Tensile Testing

Tensile testing was done on so-called dumbbell-shaped test sample. The round tensile specimen Type A (Norm, 2016) was used to determine the mechanical properties, such as Young's modulus and tensile strength, and compare theses with the properties proposed by the manufacturer. (Figure 3.12).

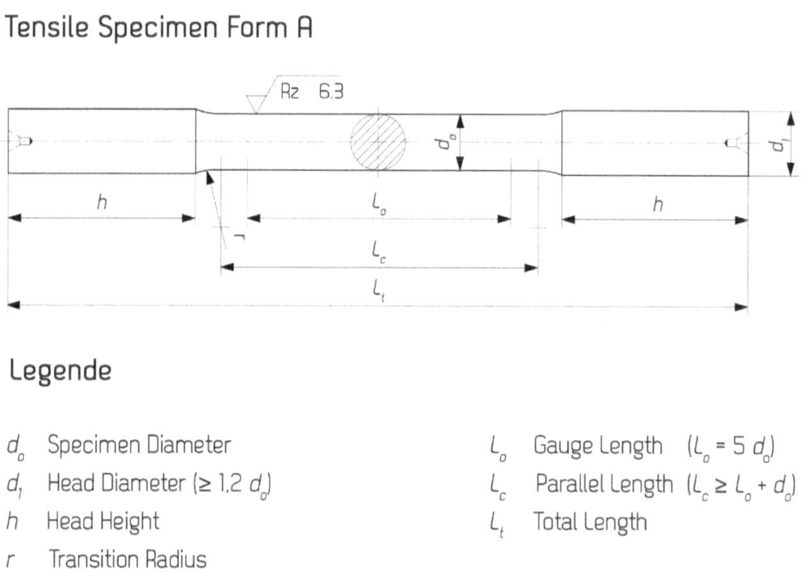

Figure 3.12 Tensile specimen Form A (DIN 50125).

Tensile testing of Stainless-steel 316L (1.4404)

The tensile testing of Stainless-steel 316L was done by the Institut für Produktionsmanagment Technologie und Werkzeugmaschinen at the Technical University Darmstadt, and the test results were provided and published at Solid Freeform Fabrication Symposium 2018 (Kniepkamp, Harbig, Seyfert, & Abele, 2018).

The round tensile specimens Type A (Norm, 2016) had a 5 mm diameter, and a 25 mm gage length. To prevent the influence of surface roughness, the tensile specimens were milled out of 10 mm diameter and 80 mm height cylinders. The specimens were tested in the build condition without heat treatment, and were printed vertically 0°, and tilted 45°. (Kniepkamp et al., 2018).

The mechanical properties were determined through two main steps. The first step was to measure the density of twenty-five of 10 mm cubes. These cubes were equally distributed on the build space. The second step was to tensile test 10 specimens that are built vertically 0°, and with an inclination of 45° each. The testing machine was Zwick Z100 (Zwick GmbH & Co. KG, Germany) (see Appendix B) and the optical videoXtens system measured the elongation with 0.5 μm resolution. The testing results (Table 3.3) laid within the range of the given values from the system manufacturer (see Appendix C) (Kniepkamp et al., 2018).

Table 3.3 Mechanical Properties of Stainless-steel 316L (1.4404).

	Density [g/cm³]	UTS 0* [MPa]	UTS 45 [MPa]	Rp0.2 0** [MPa]	Rp0.2 45 [MPa]	A 0*** [%]	A 45 [%]
Mean	7.9889	584.59	643.49	423.46	600.6	41.46	42.5
Std. dev	0.0023	0.92	1.4	9.56	11.82	0.94	0.57

* ultimate tensile strength, ** yield strength at 0.2%, *** elongation at breakpoint

Tensile testing of Aluminum ALSi10 Mg

For the tensile testing of Aluminum AlSi10 Mg, the round tensile specimens (Type A) were used one more time for the test. To prevent the influence of the surface roughness on the results, the tensile specimens were milled out of 10 mm diameter and 80 mm height cylinders. The specimens were tested in the build condition without heat treatment, and were printed vertically 0° (S), and tilted 45° (45) (Figure 3.13). Fourteen specimens were tested for each direction. The testing machine was Zwick Z100 (see Appendix B) (Zwick GbmH & Co.KG, Germany).

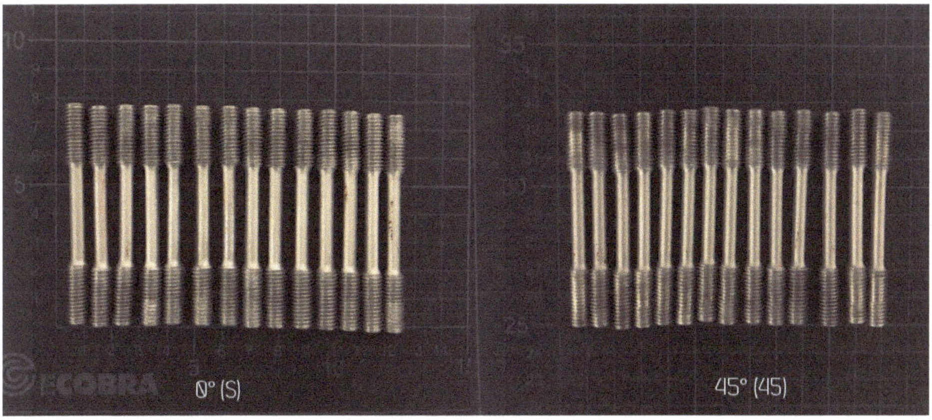

Figure 3.13 Tensile specimens; Vertically printed 0°(left)-Tilted printed 45° (right).

Each specimen was measured at three locations to determine the average diameter (Figure 3.14).

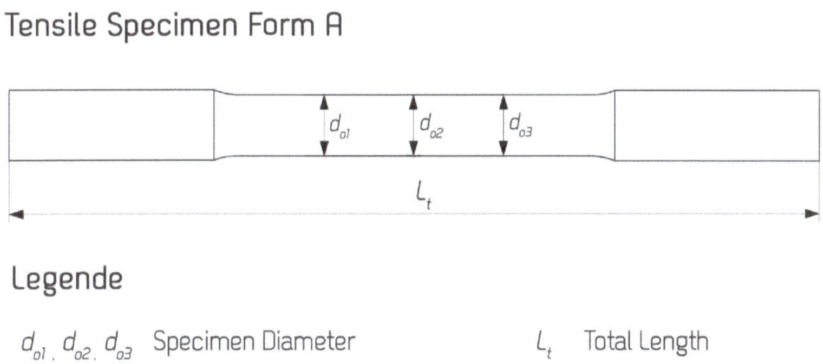

Figure 3.14 Measurements' location.

The physical measurements of the specimens that were vertically printed at 0° (S), showed slight discrepancies. However, the discrepancies laid within the manufacturing tolerance range of 3D printed elements (Table 3.4).

Table 3.4 Tensile specimens' measurements and F max in (kN) for 0° (S) specimens.

Tensile specimen	d_{01}(mm)	d_{02}(mm)	d_{03}(mm)	d_m(mm)	L_t(mm)	F max (kN)
S1	4,9	4,87	4,85	4,87	74,14	5764,49
S2	4,95	5	4,98	4,97	74,04	6144,64
S3	5	4,99	4,98	4,99	74,23	5990,84
S4	4,87	4,86	4,87	4,87	73,99	5389,70
S5	4,88	4,88	4,87	4,88	74,02	5413,61
S6	4,96	4,96	4,95	4,96	74,22	6004,38
S7	4,87	4,88	4,87	4,87	74,17	5890,88
S8	4,86	4,87	4,87	4,87	74,3	5880,64
S9	4,87	4,88	4,86	4,87	74,1	5643,00
S10	4,86	4,86	4,86	4,86	74,28	6000,65
S11	4,96	4,97	4,97	4,97	74,27	5891,48
S12	4,88	4,87	4,87	4,87	74,08	5444,74
S13	4,86	4,88	4,86	4,87	74,13	5669,60
S14	5,09	5,1	5,1	5,09	74,13	6412,67
Ø*						5750,02
Max.						6412,67
Min.						5389,70

* S2, S6, S10, & S11 are excluded from the mean calculations

As can be clearly seen in Figure 3.15, a very soft initial stiffness can be seen with two samples. Since there were problems with the fixation of these two samples, they were not considered for all further calculations.

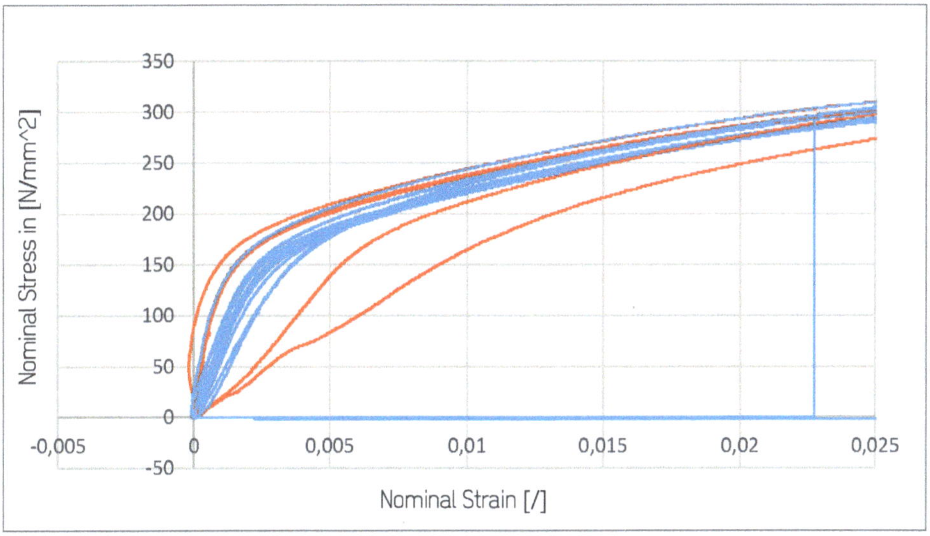

Figure 3.15 Stress strain diagram for the tested Tensile specimens, Blue curves represent the specimens included in the mean calculations – Red curves were excluded.

The visual analysis (Figure 3.16) was done using the microscope from Keyence VHX-600D. One picture was taken of the failure position at each specimen. The breakage at the failure positions showed a slight difference but a similar failure pattern.

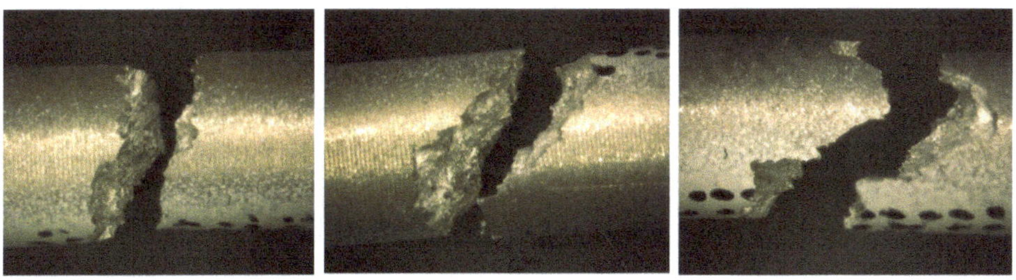

Figure 3.16 Failure positions of S01, S02 & S03 respectively.

The physical measurements of the specimens that were vertically printed at 45° (45), showed slight discrepancies. However, the discrepancies laid within the manufacturing tolerance range of 3D printed elements (Table 3.5).

Table 3.5 Tensile specimens' measurements and F max in (kN) for 45° (45) specimens.

Tensile specimen	d_{01}(mm)	d_{02}(mm)	d_{03}(mm)	d_m(mm)	L_t(mm)	F max (kN)
45-01	5,04	5,05	5,05	5,05	74,16	6025,25
45-02	5,04	5,04	5,08	5,05	74,22	6295,50
45-03	5,05	5,05	5,07	5,06	74,16	6080,95
45-04	5,07	5,04	5,06	5,06	74,15	6241,67
45-05	5,06	5,05	5,05	5,05	74,2	6010,60
45-06	5,06	5,04	5,06	5,05	74,05	6165,46
45-07	5,03	5,05	5,04	5,04	74,1	6030,57
45-08	5,04	5,04	5,07	5,05	74,19	5941,10
45-09	5,04	5,05	5,05	5,05	74,05	6050,73
45-10	5,04	5,04	5,05	5,04	74,23	6116,80
45-11	5,05	5,05	5,06	5,05	74,14	6017,33
45-12	5,04	5,05	5,05	5,05	74,23	6090,71
45-13	5,04	5,05	5,05	5,05	74,1	6063,36
45-14	5,03	5,04	5,06	5,04	74,11	6089,75
Ø*						6075,80
Max.						6295,50
Min.						5941,10

* 45-3,45-4, & 45-13 are excluded from the mean calculations

Also for the tests at a print direction of 45° there were some difficulties with the load application or the clamping of the specimens in the testing machine, so that these tests are not considered in the following calculations.

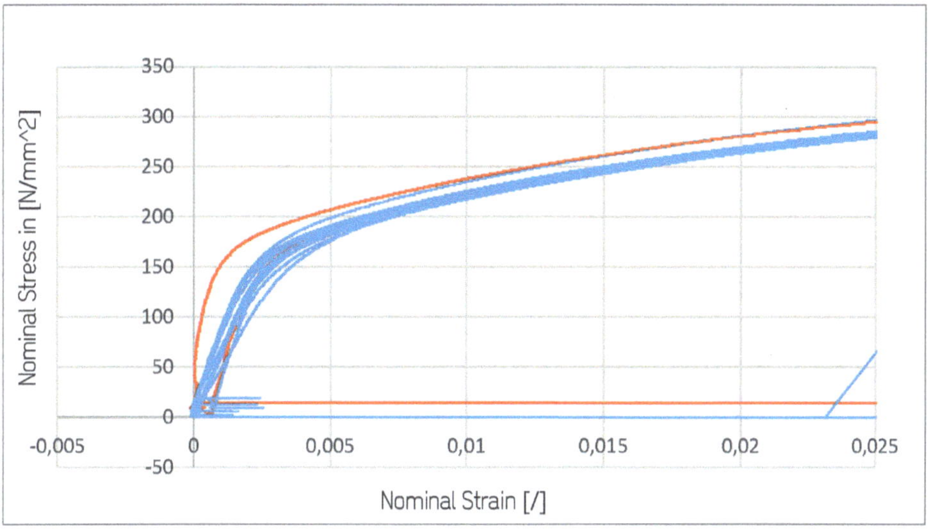

Figure 3.17 Stress diagram 2 for the tested tensile specimens, Blue curves represent the specimens included in the mean calculations – Red curves were excluded.

Visual analysis was done using the microscope from Keyence VHX-600D. One picture was taken of the failure position at each specimen (Figure 3.18). The breakage at the failure positions showed a slight difference but a similar failure pattern.

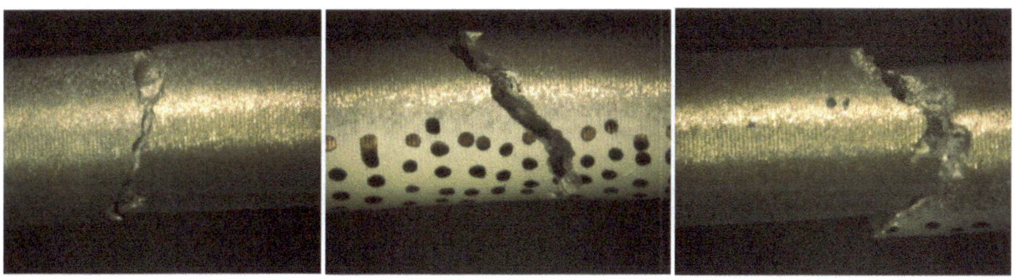

Figure 3.18 Failure positions of 45_01, 45_02 & 45_03 respectively.

The average calculations (Table 3.6) laid within the range of the given values from the system manufacturer (Appendix G).

Table 3.6 Mechanical Properties of Aluminum (AlSi10 Mg).

	Density**** [g/cm³]	UTS 0* [MPa]	UTS 45 [MPa]	Rp0.2 0** [MPa]	Rp0.2 45 [MPa]	A 0*** [%]	A 45 [%]
Mean	2,7	304	303,55	181,23	182,86	2,4	3,2
Std. dev	--	10,07	4,33	7,2	5,29	0,29	0,29

* ultimate tensile strength, ** yield strength at 0.2%, *** elongation at break point, **** from manufacturer

The Engineering of Validation test

To design an engineering validation test, several parameters had to be defined to structure the process. The parameters affect the geometry generation, the testing environment, and the building process of the validation specimens. The process is done using CAD software such as Rhinoceros, and FEA software such as ANSYS. For the testing, the universal test machine Zwick Z05 (see Appendix B) is used. The different designs and test stages will be described in the following sections.

Form-finding of the Validation Model

The parameters of the geometry are defined with respect to the limitations of the available tools. The testing machine Zwick Z050 limits the width of testing specimens to 6 cm because of the size of the clamps, and can generate a maximum tension force of 50 kN. Therefore, the validation model has to reach failure before reaching the machine's maximum loading limit.

The geometry should fulfill the criteria that request the occurrence of failure at predefined positions on the specimen. In the following, a new type of test sample will be derived which is unaxially stressed at first glance. However, the geometry is designed in such a way, that at a particular position higher-order stress state (not only uniaxial) is present. This will be proven by numerically evaluate the so-called triaxiality, which is defined as the quotient of hydrostatic stress and von Mises equivalent stress. The aim of this validation experiment is to generate the highest possible triaxiality in order to check whether the elasticity constants obtained from the uniaxial tensile tests can also describe the multi-axial stress state well.

It is common to determine the structural integrity of metal components by the comparison of σ_{VM} to the yield stress of the material. Usually, the yield stress is determined by a uniaxial test. This assumes that the defined limits are applicable to evaluate arbitrary stress conditions that would be present in complex geometries. However, if a pure shear condition is present as in a torsional test, the material yield stress would differ from the uniaxial yield stress from the tension test. In order to evaluate arbitrary stress states in complex geometries, the triaxiality η_{mod} is introduced. The triaxiality is defined by the quotient between the hydrostatic stress and the von Mises equivalent stress.

The principle Cauchy stress tensor was built parametrically to permit the illustration of the triaxiality in the three-dimensional space regarding the principal Cauchy stress tensor, it is defined as follows:

$$\boldsymbol{\sigma} = \begin{pmatrix} \sigma_{11} & 0 & 0 \\ 0 & \alpha\sigma_{11} & 0 \\ 0 & 0 & \beta\sigma_{11} \end{pmatrix} \tag{3.5}$$

From a physical viewpoint, limit values were introduced for both parameters ($-1 \leq \alpha, \beta \leq \pm 1$). The general form of triaxiality in the principal stress space results in:

$$\eta_{mod} = \left| -3\frac{\sigma_{HT}}{\sigma_{vm}} \right| = \left| -3 \frac{\frac{1}{3}(\sigma_{11}+\sigma_{22}+\sigma_{33})}{\frac{1}{\sqrt{2}}\sqrt{(\sigma_{11}-\sigma_{22})^2+(\sigma_{22}-\sigma_{33})^2+6(\sigma_{12}^2+\sigma_{23}^2+\sigma_{31}^2)+(\sigma_{33}-\sigma_{11})^2}} \right| \tag{3.6}$$

Accordingly, the amount of triaxiality is defined by the parameters α, and β. In order to display the triaxiality visually, the triaxiality must be plotted according to the parameters alpha and beta. Figure 3.19 shows this three-dimensional plot, in which the different stress states can be seen. If $\alpha = \beta = 1$, pure dilatational deformation occurs (Drass et al., 2018).

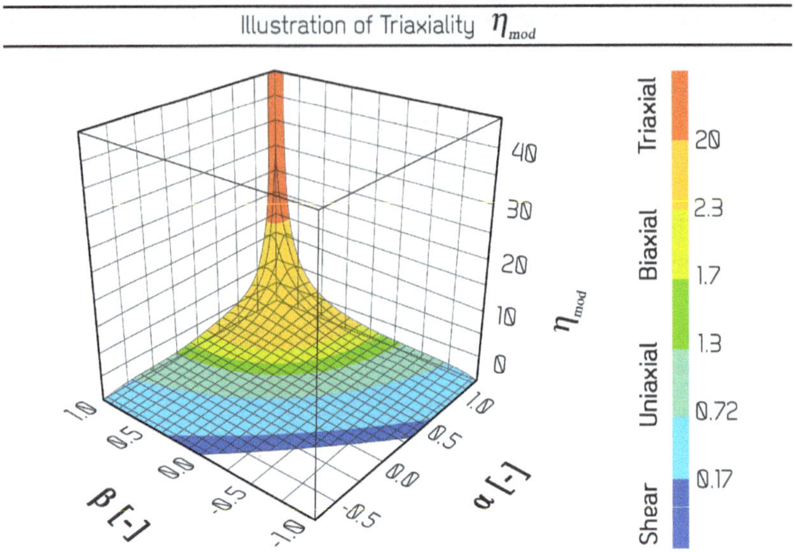

Figure 3.19 Illustration of Triaxiality (Drass et al., 2018).

Theoretical values and threshold values for triaxiality are introduced, therefore, defining a minimum and a maximum value by calculating different global stress conditions. These values are the boundary values for the calculation process (Table 3.7) (Drass et al., 2018).

Table 3.7 Theoretical values and threshold values for triaxiality.

Stress rate	**Theoretical value**	**Triaxiality mod. deviation of 10%**
Triaxial	$\eta_{Tri} = \infty$	$\eta_{Tri} \geq 20$
Biaxial	$\eta_{Bi} = 2$	$2.3 \geq \eta_{Bi} \geq 1.7$
Uniaxial	$\eta_{Uni} = 1$	$1.3 \geq \eta_{Uni} \geq 0.72$
Shear	$\eta_{Shear} = 0$	$0.17 \geq \eta_{Shear} \geq 0$

These values can be calculated by applying the routine below in ANSYS software, and the results are be the different stress conditions within the test model. With the support of these threshold values, it is possible to predict the occurrence of multiaxial stress conditions at the desired positions (Drass et al., 2018). In order to illustrate the triaxiality in ANSYS software, the following code can be directly applied as so-called user-defined result to show the triaxiality.

```
SUBROUTINE Triaxiality (S1, S2, S4)
n = abs (2 * (S1+S2+S3)
& / (6 * ((S1) ** 2 - S1 * S2 – S1 * S3
& + (S2) ** 2-S2 * S3+S3 ** 2) **0.5)) * 3
END SUBROUTINE
```

The geometry of the validation model is developed using Rhinoceros software. The model is then simulated and tested using ANSYS software. The process is repeated to reach an applicable validation model and to maximize the triaxiality occurring at the center of the test sample. Symmetry conditions were respected to save calculation time. The material model is defined in Ansys using the information that came out of the tensile testing. The idea was to use the weakest orientation of the tensile specimens, and then apply it to all directions in the material, creating an isotropic material model.

For the material model of stainless-steel, there was no complete measurement of the specimens available. Therefore, the values of the material's data sheet were used, and the slope after reaching the yield point with 0.1 of the E-modules was utilized (Karthik Alagarsamy, 2016; Shankar & Mayuram, 2008). This affects the accuracy of the results, but it was sufficient to test the hypothesis (Figure 3.20).

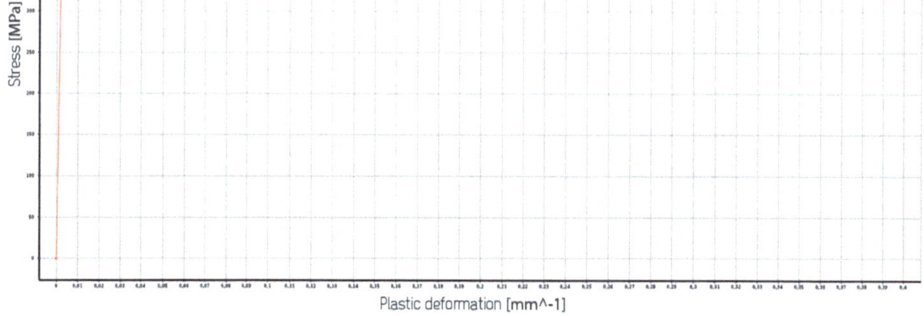

Figure 3.20 Material Model for Stainless-steel 316L (1.4404).

For the material model of aluminum, the measurement values of one of the tensile specimens were used. These values are tensile strength, yield limit at 0.2%, and Young's modulus, which lay slightly below average values. Then 30 to 40 of the exported measurement values were exported from the specimen, and the true strain and the true stress were calculated. With the E-module as a value for the elastic deformation, the material model is now defined (Figure 3.21).

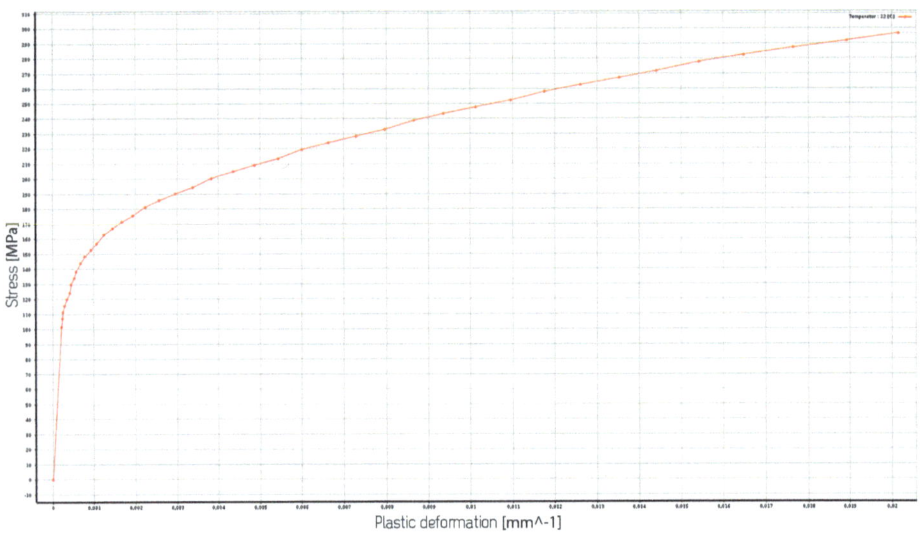

Figure 3.21 Material Model for Aluminum (ALSi10 Mg).

Next step is to define the boundary conditions of the simulations. These conditions are the definition of the support points and the direction of their deformation. One support is restricted and defined as a rigid connection in X-, Y-, Z-direction. The other side is the movable support, and where the resultant force is measured (Figure 3.22). The boundary conditions of the simulation were the same for Stainless-steel 316 L and Aluminum ALSi10 Mg. Additionally, only a half model was calculated under exploiting symmetry boundary conditions to reduce calculation time.

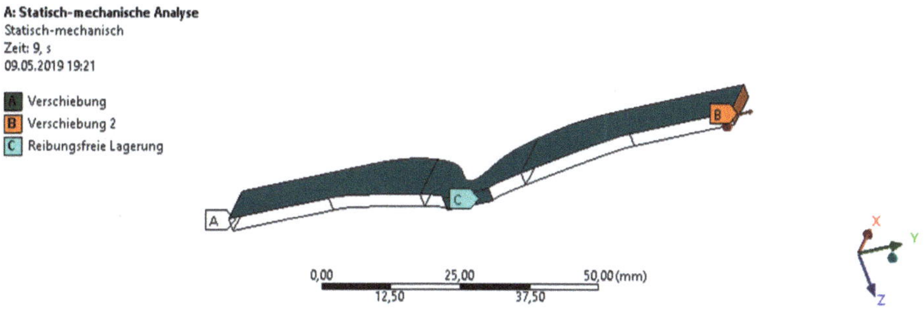

Figure 3.22 The boundary conditions for the simulation. A is the fixed support point; B is the movable, and C is defined as a movable support point.

The goal is to simplify the simulation process for the 3D printed objects and to provide a simulation model that can be trusted to avoid extensive testing.

The first version of the validation model considered for manufacture in Stainless-Steel 316L (1.4404), was a Y form. This choice was driven by the assumption that if the Y form sample is subject to tensile forces, multiaxial stresses will be present at the forking point. The shape was 97 mm in height, 45 mm in width at the two side ends, 15 mm in width at the opposite end, and 5 mm in thickness (Figure 3.23).

The triaxiality (η_{mod} = 3,88385) was sufficient based on the threshold limit value (Table 3.7) to promote the occurrence of plastic strain at the forking point of the validation model. However, it seemed that any transition of the curvature at the forking point will lead to the change of the plastic strain position (Figure 3.24).

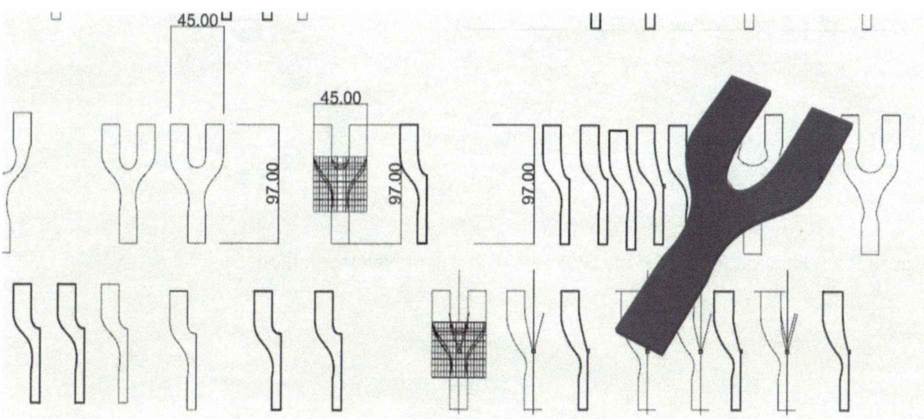

Figure 3.23 Form-finding of the validation model – early forms.

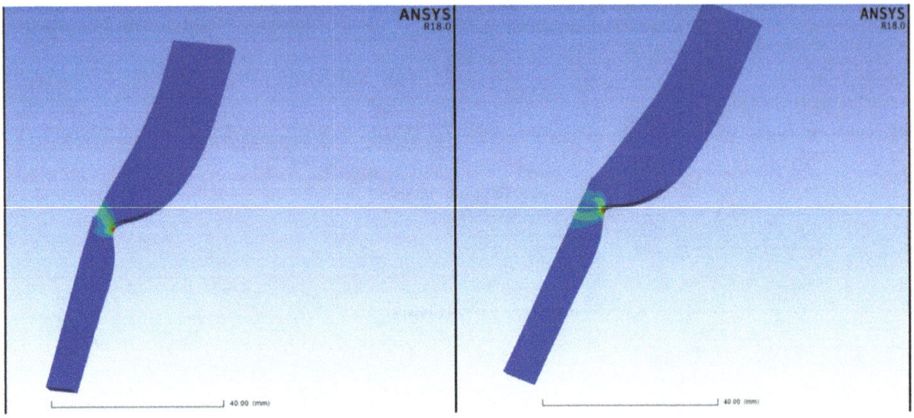

Figure 3.24 Failure location in Y-arm (left), Failure location in the Y handle (right).

The second version of the validation model was an X form. The shape was 100 mm in height, 40 mm in width, and 5 mm in thickness (Figure 3.25). The new form introduced a weaker Y direction, which induced the plastic strain to occur at the forking point, and provided a good value of triaxiality (Figure 3.26).

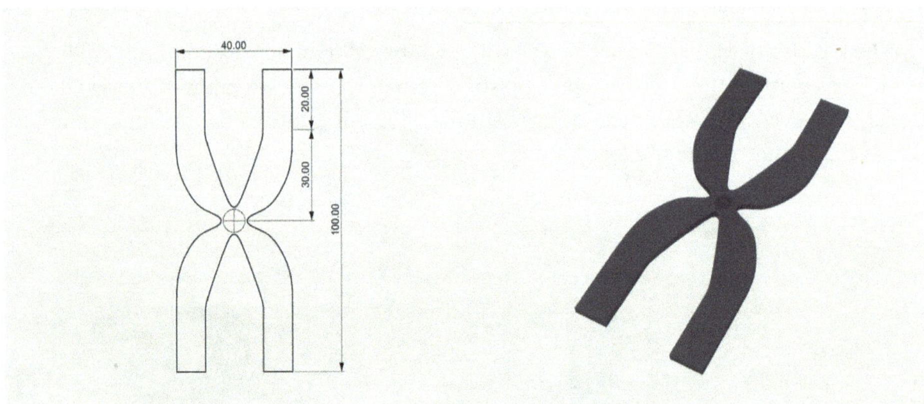

Figure 3.25 Final geometry for the Stainless-Steel 316L validation Models.

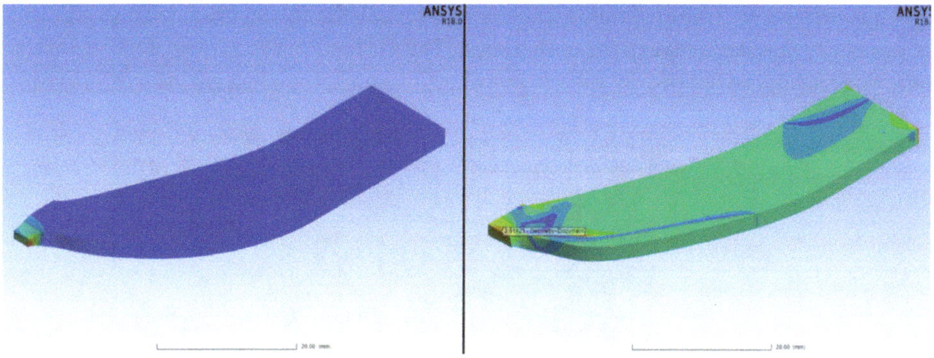

Figure 3.26 Plastic deformation at X-axis(left), Geometry triaxiality (right).

The form also added a new symmetry condition to the model that could be applied in ANSYS, and simulates ½ arm of the model. Thus, it permits the relocation of the failure position to the exact middle intersection point of the four arms. It became clear that the failure location is in the middle of the validation model. However, the failure begins at the border of the model, not from the center (Figure 3.27). Accordingly, the last modification was to add half ball recess to the middle of the X form (Figure 3.25). The third symmetry plane makes the model even smaller, which allows simulation at a finer level. ANSYS simulates an approximation of the real loading situation, and the finer the calculation network, the more accurate the results.

The equivalent plastic strain shows the amount of permanent strain in the model (Figure 3.27).

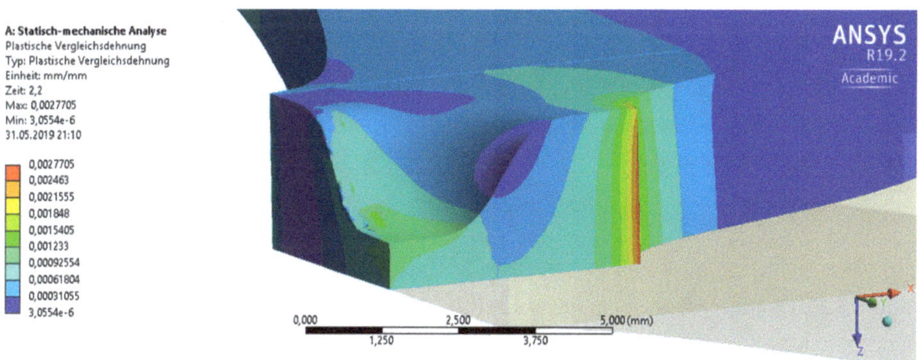

Figure 3.27 The equivalent plastic strain at 4 kN - the beginning of the failure position in the validation models.

According to the threshold of triaxiality (Table 3.7), the model provided a good value of triaxiality that lies between the occurrence of biaxial and triaxial stresses in the validation models (Figure 3.28).

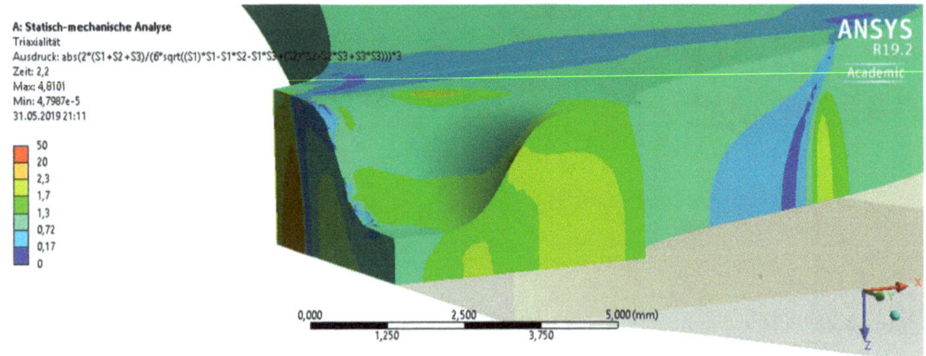

Figure 3.28 Triaxiality at 4kN – to be compared with the illustration of Triaxiality (Figure 3.20).

The same form was used to manufacture the validation model in aluminum (AlSi10 Mg) (Figure 3.29). The ball recess was removed because of post-processing complications that were noticed with the stainless-steel models. These complications will be explained in the following sections.

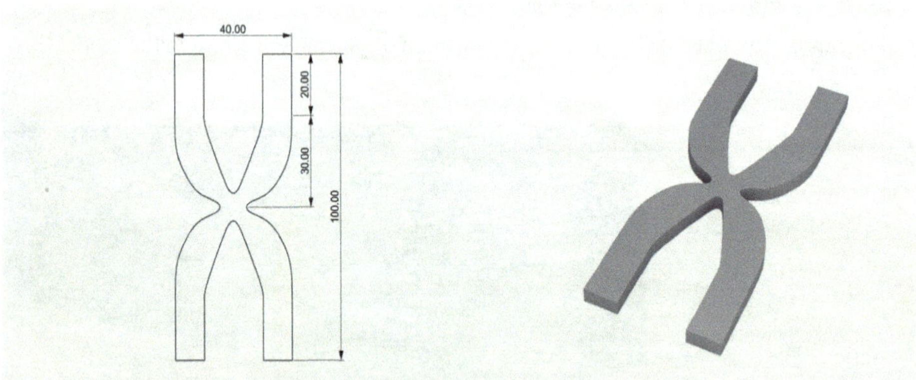

Figure 3.29 Final geometry for the Aluminum (AlSi10 Mg) validation Models.

The material model based on the tensile testing was defined, and the model was simulated. The results showed no huge effect of removing the ball recess of the models. The failure position starts at the corner of the model (Figure 3.30).

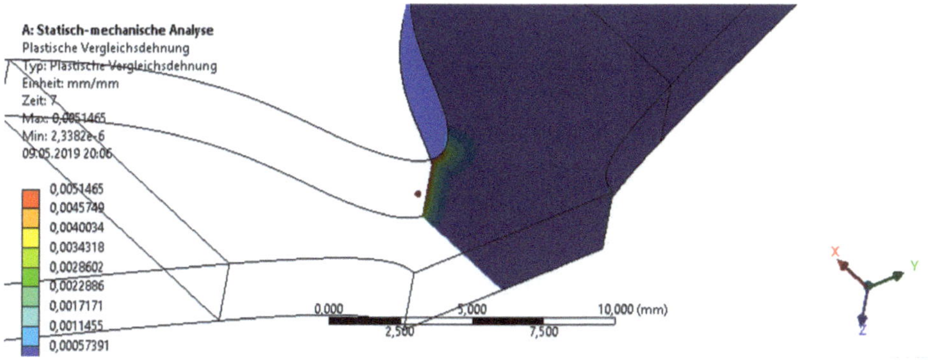

Figure 3.30 The equivalent plastic strain at 4 kN- the beginning of the failure position in the validation models.

According to the threshold of triaxiality (Table 1.7), the model provided a good value of triaxiality that lies between the occurrence of biaxial and triaxial stresses in the validation models (Figure 3.31).

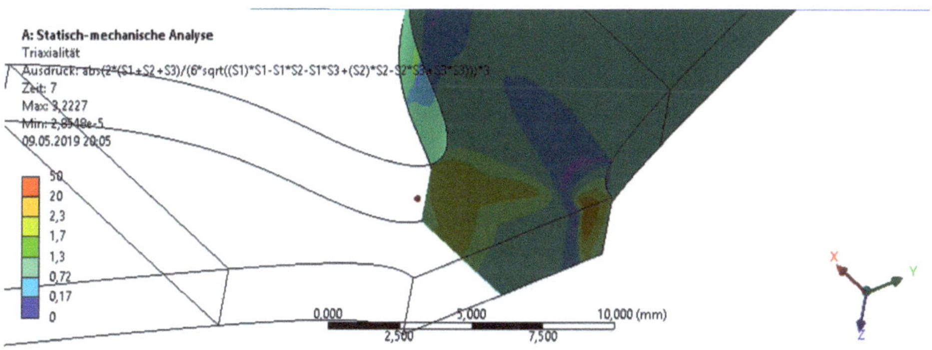

Figure 3.31 Triaxiality at 4kN – to be compared with the illustration of Triaxiality (Figure 3.20).

Manufacturing of the validation models

The Stainless-steel 316L (1.4404) models were manufactured and post-processed at Technical University Darmstadt. The print was planned for four printing orientations: horizontal 90°, vertical 0°, flat 180°, and tilted 45° (Figure 3.32).

Figure 3.32 The different print orientation of the Stainless-steel 316L (1.4404) validation model (from left to right: horizontal 90°, vertical 0°, flat 180°, tilted 45°).

The first prediction was that the flat printed specimens would present the highest load bearing capacity, and the vertical printed specimens will present the lowest load bearing capacity. The horizontally oriented specimens 90° were printed in 26 h 40 min with approximately 2148 layers and a total print height of 42.96 mm. The flat oriented specimens 180° were printed in 15 h 21 min with around 398 layers, and a total print height of 7.96 mm. The vertical 0° and the tilted 45° were not printed as separate print jobs. That made the determination of individual print time impossible. However, the entire print lasted for 69 h 33 min and generated 21 pieces. Six vertical oriented parts with print height of 103.96 mm, and six 45° tilted parts with print height of 102.96 mm belonged to this test. The job quality report of the horizontal and the flat oriented specimens is available (see Appendix D). However, for the other two directions, the job quality report was only available for the entire print job, which included irrelevant parts that printed during the same print cycle (see Appendix E).

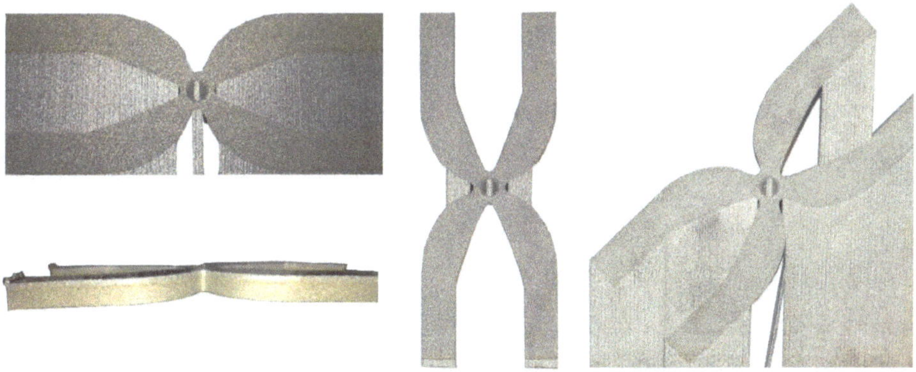

Figure 3.33 The support structure needed for each printing orientation.

Post-processing is the final step to finalize the 3D printed element and to prepare it for use. The process needs to be done carefully to avoid any damage to the element. These processes vary between cleaning, sanding, polishing, coating, heat treatment, and removing support structure. The support structure is a fine, hollow structure that builds via the melt application on thin melting paths. It helps to support the overhangs of the printed element. The support structure is cut first with an electric saw machine. Then a suitable power rotary tool is used to sand and remove the last parts of the support structure. For this experiment, a Dremel 4000-6/128 machine was used, with different heads that are suitable to machine metal parts (Figure 3.33 & 3.34).

Figure 3.34 Some of the specimens after removing the support structure.

It was evident that there were small discrepancies between the models, caused by manual postprocessing. However, based on the nature of the support structure – which is a hollow, thin grid structure – these discrepancies were disregarded.

The Aluminum (AlSi10 Mg) models were manufactured by a 3D printing service provider. The print was planned for four printing orientations: horizontal 90°, vertical 0°, flat 180°, and tilted 45° (Figure 3.35).

Figure 3.35 The different print orientation of the Aluminum (AlSi10 Mg) validation model (from left to right: horizontal 90°, vertical 0°, flat 180°, tilted 45°).

The models were in perfect condition and had been professionally post-processed. Accordingly, the discrepancies were within the provided range from the manufacturer (Figure 3.36).

Figure 3.36 The Aluminum Validation Models.

Testing of the validation models

Figure 3.37 Test preparation.

After post-processing, the validation models were ready to be tested. For the test, a Zwick testing machine Z050 was used (see Appendix B). Different steps were needed to run the test. These steps started with spraying the validation models with a white coating, then with a black from a decent distance to induce a black speckled pattern with a white background on the models. This coloring was needed for the visual evaluation that is done later by VIC 3D. The validation model was then fixed with two pneumatic clamps connected to a digital force gauge with maximum measuring load of 50 kN. Then cameras were set up. Different calibration plates were used to focus the cameras' lenses on the fixed model (Figure 3.37). After calibration, the cameras were ready to take pictures of the model at a rate of 1 picture/second. These pictures were used later for visual evaluation. The Zwick machine put the validation model under uniaxial forces with 2 kN/min until failure. The process was repeated until all the available validation models are tested.

Evaluation of the test results

To obtain a full evaluation of the test, three criteria had to be evaluated; a) visual analysis of the failure positions and how the material behaves, b) the force measurement of the Zwick machine's digital force gauge, and c) the deformation and the strain analysis done by VIC-3D. After concluding the test results, the comparison between the results and the simulations was done to determine the accuracy of the simulations.

The visual analysis was done using the microscope from Keyence VHX-600D. Three pictures were taken from the top, bottom, and side of the failure position at each specimen.

Stainless-steel 316L (1.4404)

For the Stainless Steel 316L (1.4404), the specimens showed almost accurate breaking art as the simulations predicted (Figure 3.38, 3.39, 3.40 & 3.41). The difference in print direction did not present a huge difference in breaking behavior. However, the maximum forces that the specimens could support were not similar.

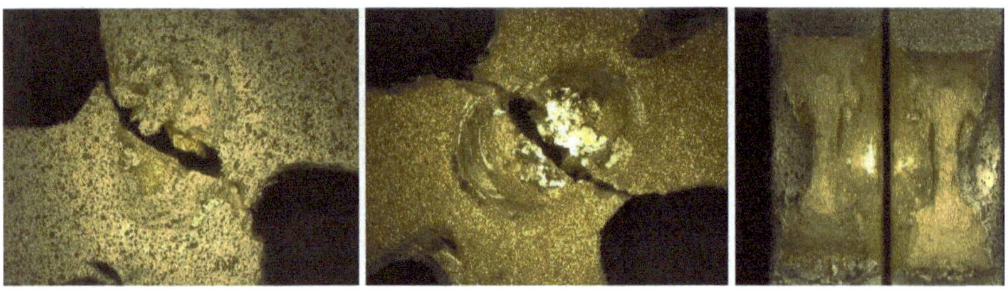

Figure 3.38 Failure positions of "horizontal 90°-04" models (front, back, & cross-section).

Figure 3.39 Failure positions of "vertical 0°-01" models (front, back, & cross-section.

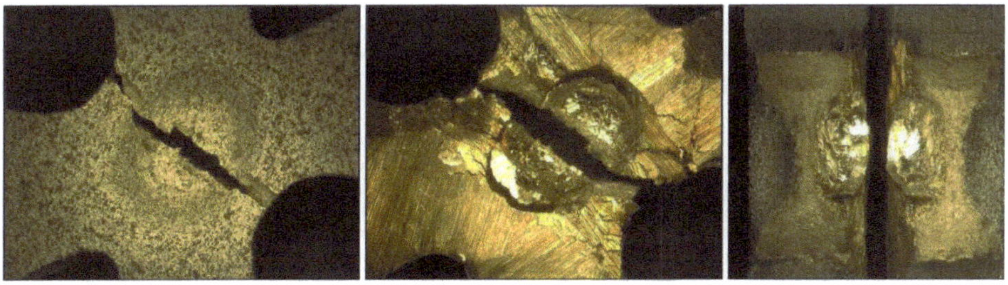

Figure 3.40 Failure positions of "flat 180°-05" models (front, back, & cross-section).

Figure 3.41 Failure positions of "tilted 45°-04" models (front, back, cross-section).

Six specimens of each printing orientation were tested. Each printing orientation presented similarity of the failure art. The overall visual analysis showed no significant change in the failure positions between all printed orientations. The breakage of the "flat 180°" at the half ball recess embraced the idea of neglecting the effect of the support structure at the recess. Accordingly, for the "vertical 0°", and "tilted 45° the support structure at the recess was not removed to avoid causing any damage to the specimens.

With its digital force gauge, the Zwick machine records the force progression along the specimens until failure. These data are presented by the maximum force that is needed to bring each specimen to the breaking point (Table 3.8, 3.9, 3.10, & 3.11).

Table 3.8 "horizontal 90°" L – notes, Max. uy, and F max. in (kN).

Validation specimen	Notes	Max. uy	F max. (kN)
L1		0.47	25.982
L2	part of the image documentation went missing	--	26.222
L3		0.54	26.319
L4		0.48	26.628
L5		0.56	26.186
L6		0.47	25.755
Ø		0.50	26.182

The horizontal 90° specimens have a minimum breaking force of 25.755 kN at the specimen L6, and a maximum breaking force of 26.628 kN at the specimen L4. The standard deviation of the force is 0.873 kN. As well, they have a minimum deformation of 0.47 at the specimen L6, and a maximum deformation of 0.74 at the specimen L1. The standard deviation of the deformation is 0.09.

Table 3.9 "vertical 0°" S – notes, Max. uy, and F max. in (kN).

Validation specimen	Notes	Max. uy	F max. (kN)
S1		0.31	24.111
S2		0.24	23.996
S3		0.30	24.082
S4		0.34	24.066
S5	The image documentation went missing	--	24.391
S6		0.28	24.440
Ø		0.29	24.181

The vertical 0° specimens had a minimum breaking force of 23.966 kN at the specimen S2, and a maximum breaking force of 24.440 kN at the specimen S6. The standard deviation of the force is 0.444 kN. As well, they have a minimum deformation of 0.24 at the specimen S2, and a maximum deformation of 0.34 at the specimen S4. The standard deviation of the deformation difference is 0.03.

Table 3.10 "flat 180°" F – notes, Max. uy, and F max. in (kN).

Validation specimen	Notes	Max. uy	F max. (kN)
F1		0.52	27.568
F2	Only one arm was under stress because of a faulty fixing condition	0.35	18.602
F3		0.48	27.884
F4	Preliminary specimen	0.51	27.427
F5		0.55	27.286
F6	One arm slipped out of the clamps during the test	0.39	24.884
Ø	Without F2 & F6	0.51	27.541

The flat 180° specimens have a minimum breaking force of 27.286 kN at the specimen F5, and a maximum breaking force of 27.568 kN at the specimen F1. The standard deviation of the force is 0.282 kN. As well, they have a minimum deformation of 0.48 at the specimen F3, and a maximum deformation of 0.55 at the specimen F5. The standard deviation of the deformation difference is 0.07.

The bottom side of the samples, was completely covered by the support structure. The manual post-processing caused unequal thicknesses, which caused the faulty measurement of specimens F2 & F6 (Figure 3.42 & 3.43).

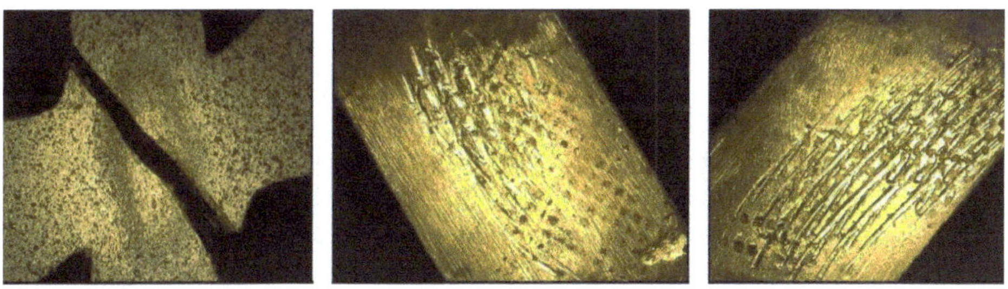

Figure 3.42 F2 specimen – faulty fixing condition.

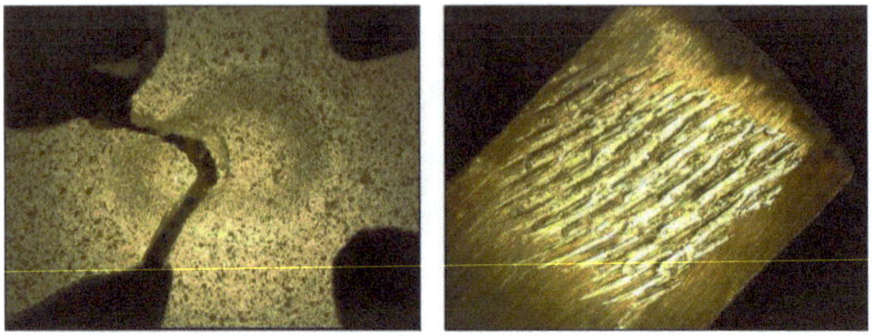

Figure 3.43 F6 specimen – faulty fixing condition

Table 3.11 "tilted 45°" 45 – notes, Max. uy, and F max. in (kN).

Validation specimen	Notes	Max. uy	F max. (kN)
45-1		0.23	25.423
45-2		0.26	25.481
45-3		0.22	25.432
45-4		0.28	25.466
45-5		0.24	25.402
45-6	The integrity of the sample was affected by damage during postprocessing	--	21.733
Ø	without 45-6	0.25	25.441

The tilted 45° specimens had a minimum breaking force of 25.402 kN at specimen 45-5, and a maximum breaking force of 25.481 kN at specimen 45-2. The standard deviation of the force difference is 0.079 kN. As well, they have a minimum deformation of 0.22 at the specimen 45-3, and a maximum deformation of 0.28 at the specimen 45-4. The standard deviation of the deformation difference is 0.02.

The specimen 45-6 was partially damaged during postprocessing, which lead to the mature breakage (Figure 3.44).

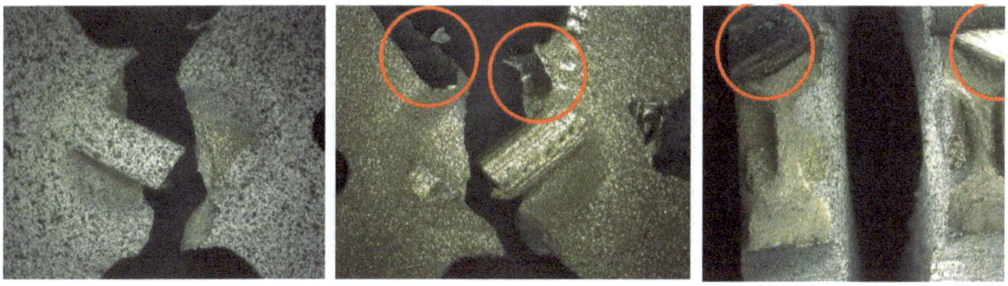

Figure 3.44 45-6 specimen – damaged during post-processing.

Next, the force-deformation curves were generated by the correlation between the measured forces out of Zwick machine, and the captured pictures of the actual deformation from the cameras. The diagrams were generated using VIC-3D, where the Y-axis is representing the force F in kN, and the X-axis is representing the deformation in Y- direction in mm.

The diagram of the force F in kN and the deformation in Y-direction in mm was generated from the simulations. The simulation does not consider the layer structure of the 3D printed elements to reduce the simulations' complexity and the assumption was made that the model is a cast object. The generated diagram is compared with all diagrams of each direction (Figure 3.45).

The results validated the assumption that was made. Despite the inaccuracy of the material model in Ansys, the physical models demonstrated better performance in comparison to the simulations. To improve the process, the material model correlates to a tensile test and that was done for the test of the aluminum material.

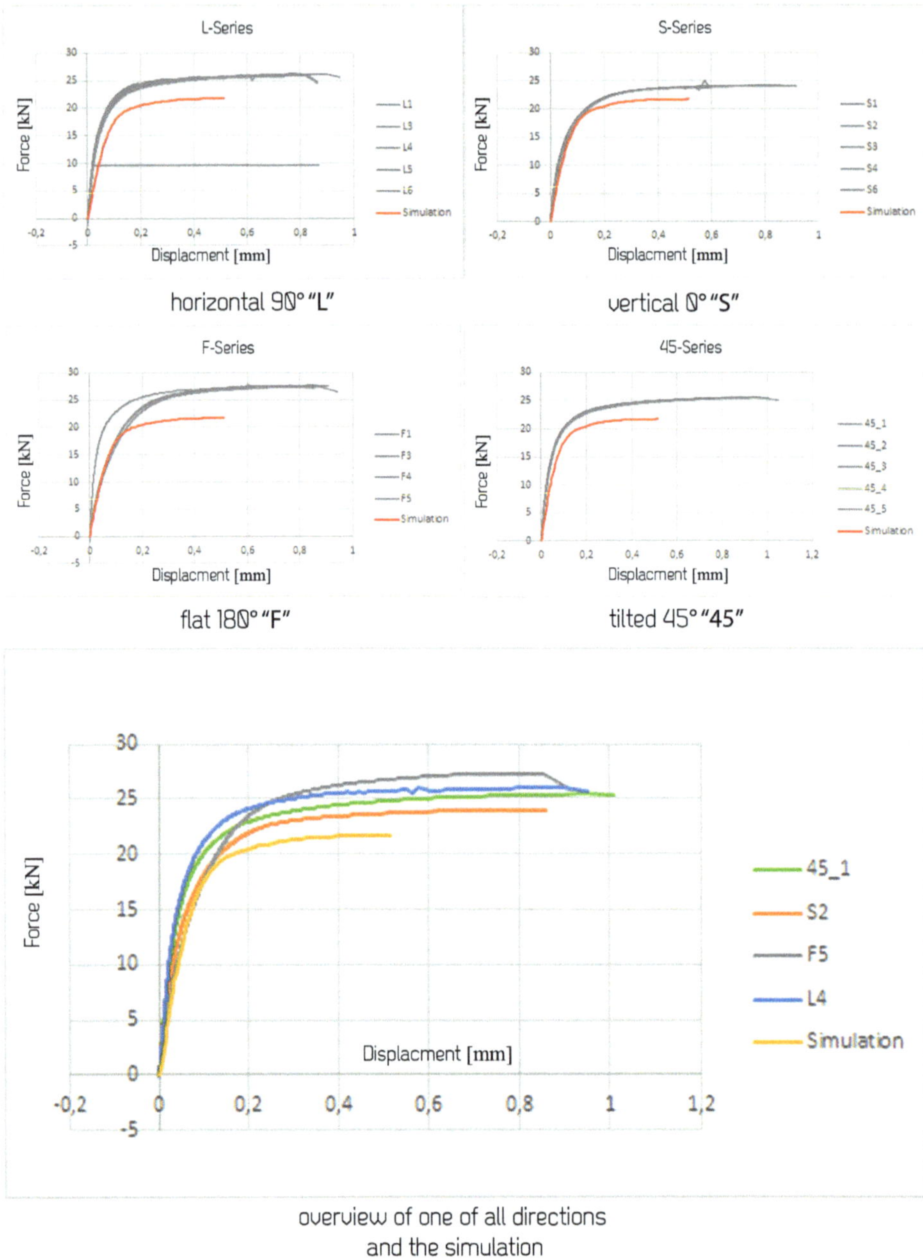

Figure 3.45 Force and deformation diagrams of all tested specimens. The simulation diagram is added in red in all charts to simplify the evaluation of the tests and in orange in the overview.

The cameras took one picture per second. These pictures were evaluated using VIC-3D software. The software compares the change of the pattern on the surface of the specimen. The area of the analysis, called "area of interest," is firstly defined in the software. AOI is what the software analyzes for deformation.

The evaluation was done for all different print orientations. For the horizontal 90° "L" L4 is presented and showed a maximum deformation of 0.009 mm at 15.6 kN and 0.022 mm at 21 kN. For the vertical 0° "S" S2 is presented and showed a maximum deformation of 0.015 mm at 15.6 kN and 0.037 mm at 21 kN. For the flat 180° "F" F5 is presented and showed a maximum deformation of 0.015 mm at 15.6 kN and 0.037 mm at 21 kN. For the tilted 45° "45" 45-01 is presented and showed a maximum deformation of 0.022 mm at 15.6 kN and 0.036 mm at 21 kN. Finally, the simulation showed a maximum deformation of 0.042 mm at 15.6 kN and 0.172 mm at 21 kN, which validate the assumption taken to simulate the 3D printed elements as casted parts (Figure 3.46) but still the difference between the simulation and the actual parts can be improved by increasing the accuracy of the material model and this is to be considered for the test of the aluminum material.

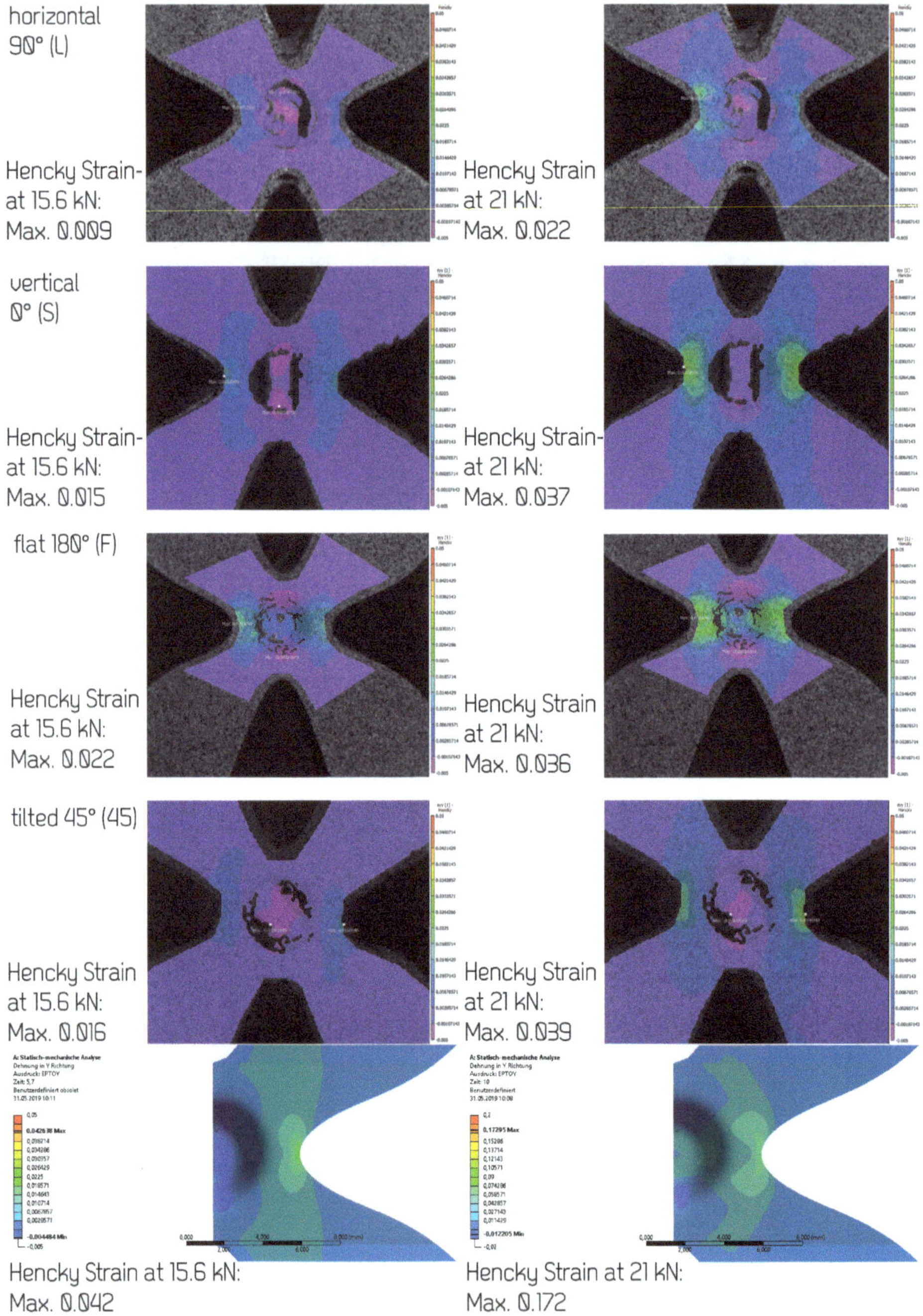

Figure 3.46 Strain analysis using Vic-3D and Ansys. The pictures show the Hencky strain of one specimen of each print orientation and the Ansys simulation (left at 15.6kN- right at 21kN).

Aluminum AlSi10 Mg

For the Aluminum (AlSi10 Mg), the specimens showed breaking art that was almost identical to those predicted by the simulations (Figure 3.47, 3.48, 3.49 & 3.50). The difference in print direction did not present a huge difference in breaking art. However, the maximum forces that the specimens could support, were not similar.

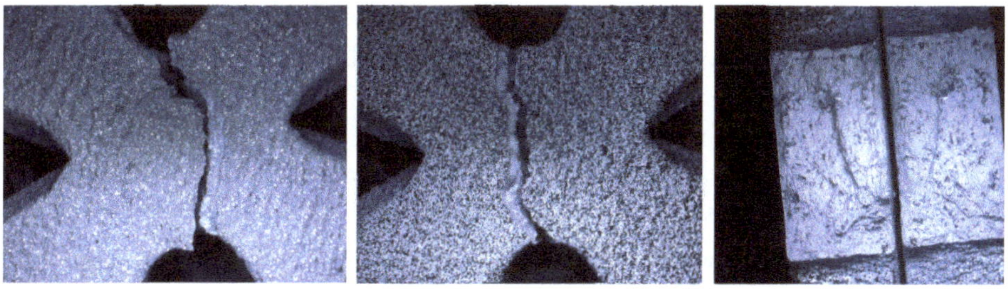

Figure 3.47 Failure positions of "horizontal 90°-05" models (front, back, & cross-section).

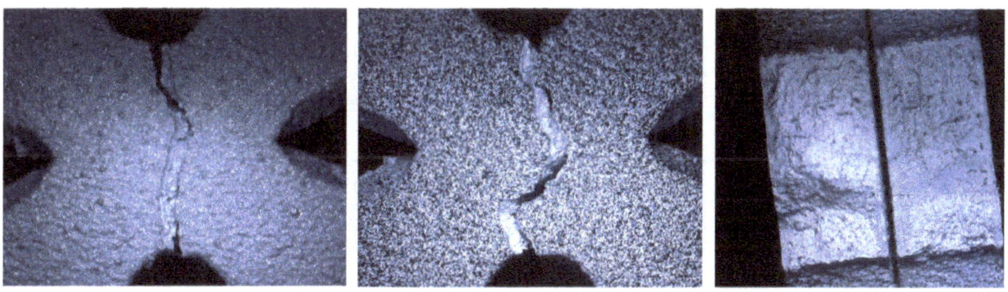

Figure 3.48 Failure positions of "vertical 0°-02" models (front, back, & cross-section).

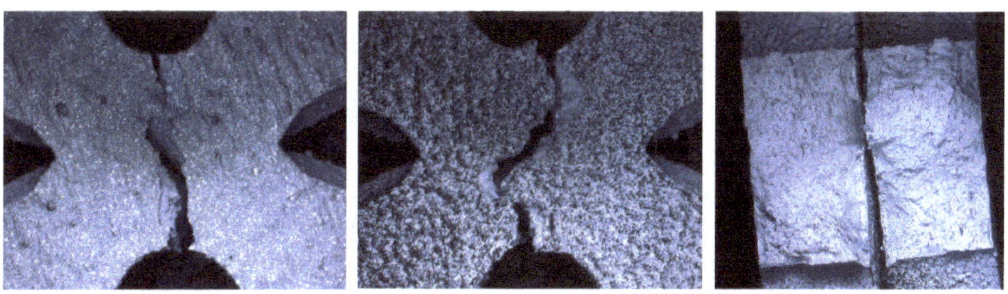

Figure 3.49 Failure positions of "flat 180°-05" models (front, back, & cross-section)

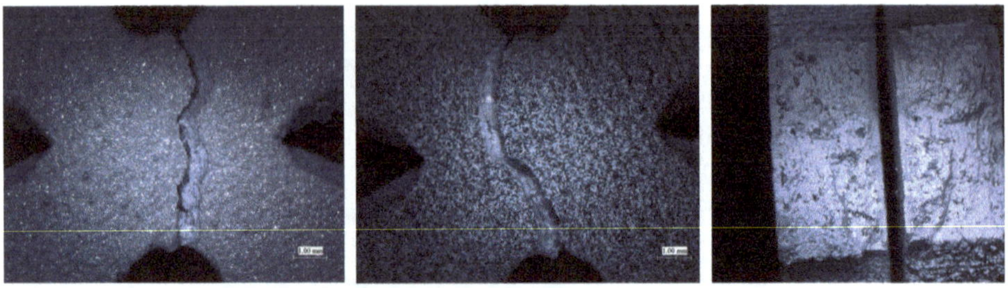

Figure 3.50 Failure positions of "tilted 45°-03" models (front, back, & cross-section)

Six specimens of each printing orientation were tested. Each printing orientation presented similar failure art. Overall visual analysis showed no significant change in the failure positions between all printed orientations.

Similar to the stainless-steel test, the Zwick machine recorded the force progression along with the specimens until failure. These data are presented by the maximum force needed to bring each specimen to the breaking point (Table 3.12, 3.13, 3.14, &3.15).

Table 3.12 "horizontal 90°" L – notes, Max. uy, and F max. in (kN).

Validation specimen	Notes	Max. uy	F max. (kN)
L1		0.036	8.14
L2		0.024	7.59
L3		0.034	8.83
L4		0.029	9.00
L5		0.047	8.32
L6		0.032	8.56
Ø		0.033	8.41

The horizontal 90° specimens have a minimum breaking force of 7.59 kN at the specimen L2, and a maximum breaking force of 9 kN at the specimen L4. The standard deviation of the force is 0.51 kN. As well, they have a minimum deformation of 0.024 at the specimen L2, and a maximum deformation of 0.047 at the specimen L5. The standard deviation of the deformation is 0.007.

Table 3.13 "vertical 0°" S – notes, Max. uy, and F max. in (kN).

Validation specimen	Notes	Max. uy	F max. (kN)
S1		0.044	10.24
S2		0.039	9.94
S3	Missing εy values	--	10.27
S4		0.062	10.16
S5		0.064	9.93
S6		0.049	9.88
Ø		0.051	10.07

The vertical 0° specimens have a minimum breaking force of 9.88 kN at the specimen S6, and a maximum breaking force of 10.27 kN at the specimen S3. The standard deviation of the force is 0.17 kN. As well, they have a minimum deformation of 0.039 at the specimen S2, and a maximum deformation of 0.064 at the specimen S5. The standard deviation of the deformation is 0.009.

Table 3.14 "flat 180°" F – notes, Max. uy, and F max. in (kN).

Validation specimen	Notes	Max. uy	F max. (kN)
F1		0.067	11.91
F2		0.051	11.44
F3		0.064	12.01
F4	Faulty fixing conditions that led to unexpected breaking behavior	--	0
F5		0.062	12.12
F6		0.051	11.45
Ø		0.059	9.82

The flat 180° specimens have a minimum breaking force of 11.44 kN at the specimen F2, and a maximum breaking force of 12.12 kN at the specimen F5. The standard deviation of the force difference is 0.29 kN. As well, they have a minimum deformation of 0.051 at the specimen F2&6, and a maximum deformation of 0.067 at the specimen F1.

The standard deviation of the deformation is 0.009.

The specimen F4 was not considered in the mean calculation because of faulty fixing conditions that lead to unexpected breaking behavior (Figure 3.51).

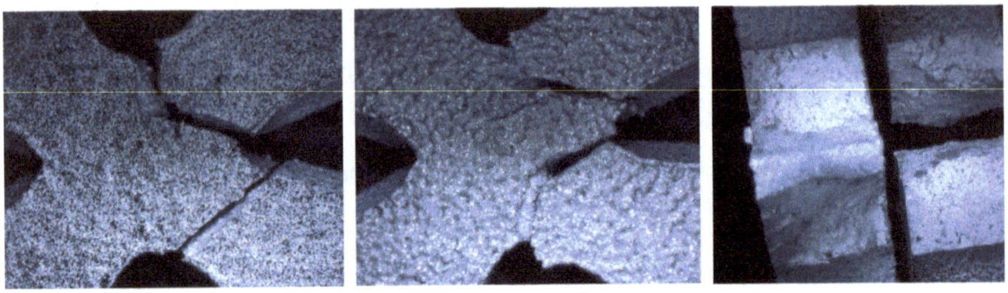

Figure 3.51 F4 specimen- Faulty fixing condition that caused strange breaking art

Table 3.15 "tilted 45°" 45 – notes, Max. uy, and F max. in (kN).

Validation specimen	Notes	Max. uy	F max. (kN)
45-1		0.045	9.29
45-2		0.034	9.26
45-3		0.030	9.46
45-4		0.029	9.80
45-5		0.031	9.02
45-6		0.035	10.31
Ø		0.034	9.52

The tilted 45° specimens have a minimum breaking force of 9.02 kN at the specimen 45-5, and a maximum breaking force of 10.31 kN at the specimen 45-6. The standard deviation of the force is 0.46 kN. As well, they have a minimum deformation of 0.029 at the specimen 45-4, and a maximum deformation of 0.045 at the specimen 45-1. The standard deviation of the deformation is 0.005.

Next, the force-deformation curves were generated by the correlation between the measured forces out of Zwick machine and the captured pictures of the actual deformation out of the cameras. The diagrams were generated using VIC-3D, where the Y-axis is representing the force F in N, and the X-axis is representing the deformation in Y- direction in mm.

The diagram of the force F in N and the deformation in Y-direction in mm was generated from the simulations. The simulation does not consider the layer structure of the

3D printed elements to reduce the simulations' complexity and the assumption was made that the model is a casted object. The generated diagram compares all diagrams of each direction (Figure 3.52).

The results validated the assumption that was taken. This assumption was to simulate the elements as a casted parts. In most cases, all directions demonstrated better behavior in comparison to the simulation.

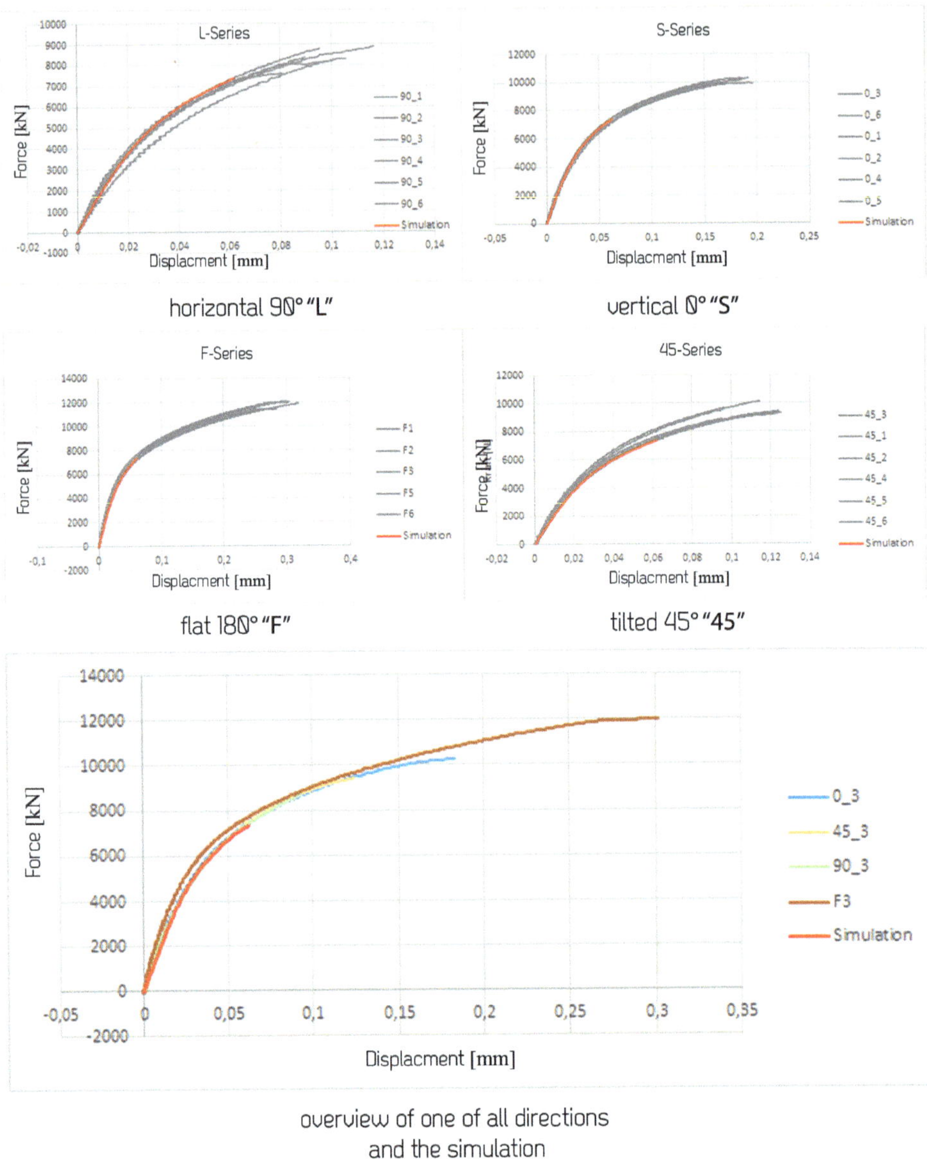

Figure 3.52 Force and deformation diagrams of all tested specimens. The simulation diagram is added in red in all charts to simplify the evaluation of the tests.

The cameras took one picture per second. These pictures were evaluated using VIC-3D software. The software compares the change of the pattern on the surface of the specimen. The area of the analysis, called "area of interest," is firstly defined in the software. AOI is what the software analyzes for deformation.

The evaluation was done for all different print orientations. For the horizontal 90° "L" L6 is presented and showed a maximum deformation of 0.006 mm at 4 kN and 0.013 mm at 6.5 kN. For the vertical 0° "S" S4 is presented and showed a maximum deformation of 0.008 mm at 4 kN and 0.014 mm at 6.5 kN. For the flat 180° "F" F1 is presented and showed a maximum deformation of 0.005 mm at 4 kN and 0.011 mm at 6.5 kN. For the tilted 45° "45" 45-03 is presented and showed a maximum deformation of 0.006 mm at 4 kN and 0.014 mm at 6.5 kN. Finally, the simulation showed a maximum deformation of 0.006 mm at 4 kN and 0.017 mm at 6.5 kN, which validate the assumption taken to simulate the 3D printed elements as casted parts (Figure 3.53).

The results proved the assumption that additive manufactured parts can be simulated as casted parts. This benefits the design process by dramatically reducing the computing time. The common simulation concept, often used to simulate additive manufactured parts, is to consider the melting process between each layer. This allows only small finite model elements to be constructed to reduce the number of elements that need to be calculated. Thus, simulating a complete element is quite a challenge. Once the isotopic behavior is neglected, the simulation process is simplified, 3D printed elements can be reliably calculated as standard elements.

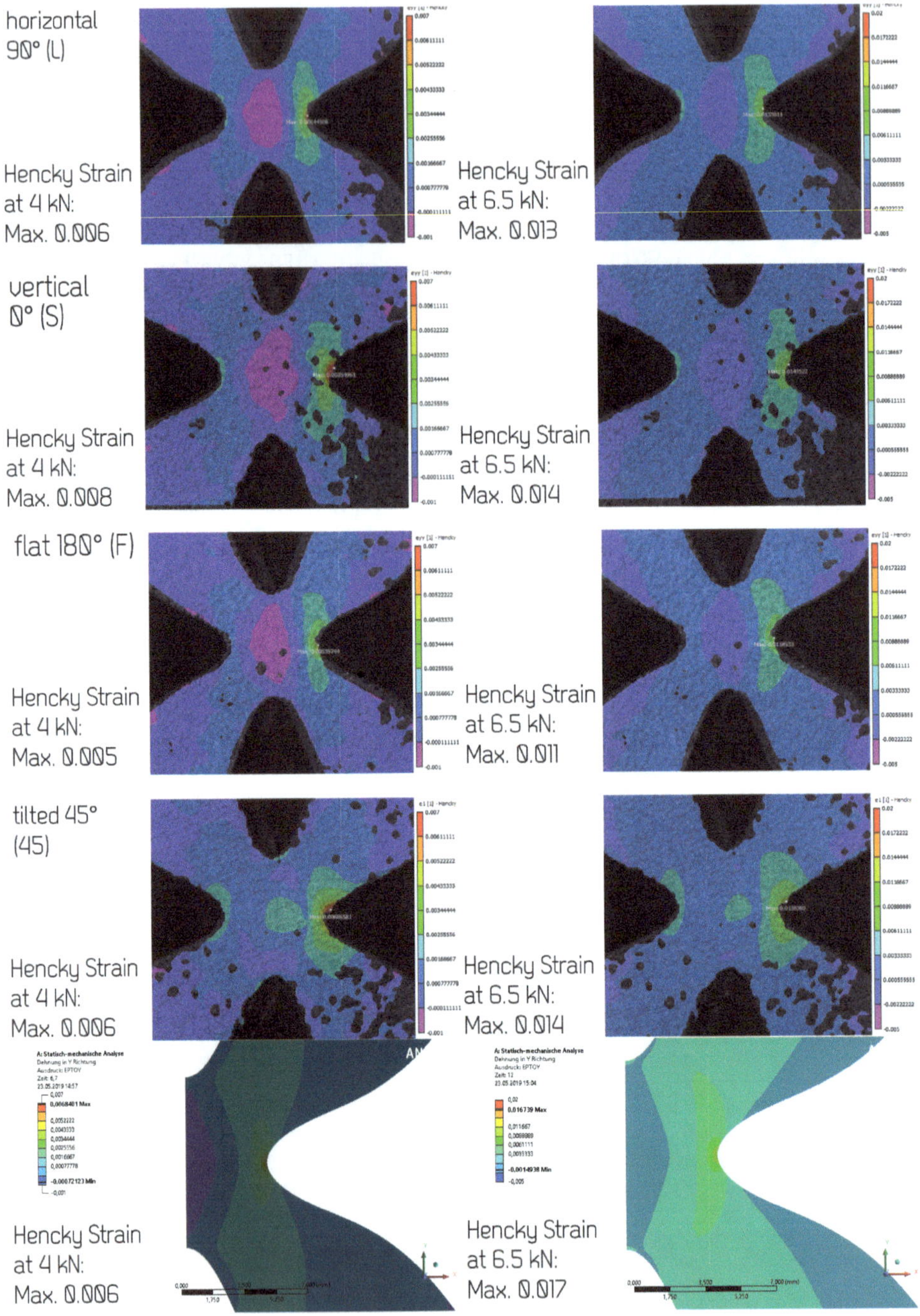

Figure 3.53 Strain analysis using Vic-3D and Ansys. The pictures show the Hencky strain of one specimen of each print orientation and the Ansys simulation (left at 4kN- right at 6.5kN).

4 Design Development

4.1 Geometric analysis

The focus on geometrical façade design will help to understand the transition from design to constructive model then to machined parts. For the realization of complex façade constructions, mullion-transom façade systems offer many design possibilities based on their dimensions and efficient manufacturing, and single-layered shell structures are often used. To clarify the transition from design to constructive model, the basics of the digital representation of free-form surfaces and its generation had to be investigated, and the NURBs-method and its parameter had to be analyzed.

4.1.1 Free-form façades/building envelopes

Façades/building envelopes are the first part of the building that a human being interacts with when they approach a building. They are a tool that represents the unique aspects of a building. Therefore, façades/building envelopes with complex geometries develop rap-idly and become more valuable. The utilization of free-form surfaces in architecture is currently defined by many terms, such as Blob-architecture, Biomorphism, and organic architecture. The development of 3D modeling software encouraged the architects to scrutinize free forms as they are in nature and move away from the standard classic forms. 3D modeling software offers the ability to visualize many kinds of surfaces, such as single curved, double curved, Convex, and Concave, digitally.

The realization of free-form surfaces is a complex process and has different approaches, such as (a) when the geometry choice is completely free, then usually a predefined rectangular gird/cell is generated to define the free-form surface; (b) when the free-form surface is generated through the translation of curves network, although this method limits the surface generation possibilities; (c) when the NURBS curves are used to form NURBS surfaces; and (d) when a triangular/quad meshing is used to form an approximation of a free-form surface. Each approach has each own advantages and disadvantages. (Schiftner, Leduc, Bompas, Baldassini, & Eigensatz, 2013).

A. Mohsen, *Design to Manufacture of Complex Building Envelopes*,
Mechanik, Werkstoffe und Konstruktion im Bauwesen 56,
https://doi.org/10.1007/978-3-658-30204-7_4

Free-form Surfaces

Façade/building envelopes are the skin of the building, represented as a surface. This surface consists of two-dimensional shapes that are parts of the three-dimensional geometric Euclidean space. Accordingly, surfaces can contain any two-dimensional objects, such as points, lines, curves, and curves network, which will affect the form and the visual appearance of the façade.

Surfaces are classified based on their area and curvature. Therefore, Gaussian curvature is one of the most important properties that define a surface. The curvature defines how much a surface deviates from a tangent surface at a specific point. The curvature is defined locally. As a result, a normal vector for each point on a surface can be generated. For a regular surface in space R^3 with a defined point on this surface, the curvature at this point can be formulated (Equation 4.1):

$$K = k_1.k_2 = \frac{1}{r_1} \cdot \frac{1}{r_2} \tag{4.1}$$

The Gaussian curvature K is the product of the principal curvature k_1 and k_2 at a given point on a specific surface. The principal curvature radius r_1 and r_2 are for both orthogonal directions. Accordingly, the curvature of a sphere is $K=1/r^2$, and for a cylinder is $K=0$. Surfaces are classified into plane (when both the principal curvature, r_1 and r_2 =0-planar-), single curved (when one of the principal curvature, r_1 or r_2 =0-parabolic-), and double curved (when the Gaussian curvature $K>0$ or $K<0$-sinclastic/anticlastic-) (Figure 4.1) (Kühnel, 2013; Sören Stephan, Sánchez Alvarez, & Knebel, 2004).

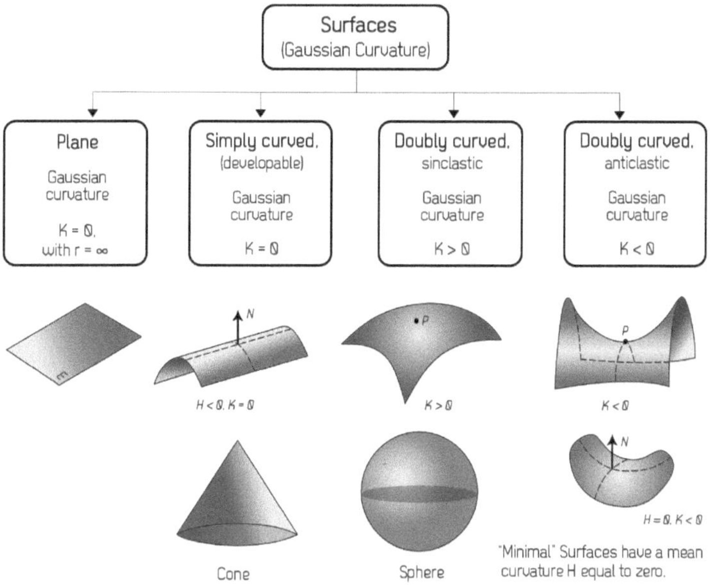

Figure 4.1 Schematic shows the classification of surfaces based on its curvature (Soeren Stephan, Sánchez-Alvarez, & Knebel, 2004).

The generation of free-form surfaces can occur as a geometric mode of generation, as well as a non-geometric mode of generation that requires a form-giving agent.

Through the non-geometric generation process, which is called form-finding, the form is generated based on the deformation caused by a certain force (shaping force) on an arbitrary surface that results in an equilibrium surface. The support points at the corners of a surface affect the generated end form. The end form of the surface is subject to physical principles that result in a homogenous force flow and internal force distribution. The equilibrium surface develops a membrane stress state – as in soap skin (minimal surface) – that only normal forces will develop, and no bending moment will occur.

Through the geometric generation process, the surface is generated based on straight and curved lines. The translation, stretching, or rotation of these lines generates the surface. The surfaces can be classified into one of the following: a rotation surface, which is when a curve is rotated around a defined rotation axis, and is segmented radially, a translation surface, which is when the translation is guided through the curvature of the guide lines, or a ruled surface, which is when straight lines move solely to create a flat surface that is parallel to these lines (Figure 4.2). The geometrical and mechanical properties of the generated surfaces are related to the form of the guidelines (Kühnel, 2013; Stahr, 2008).

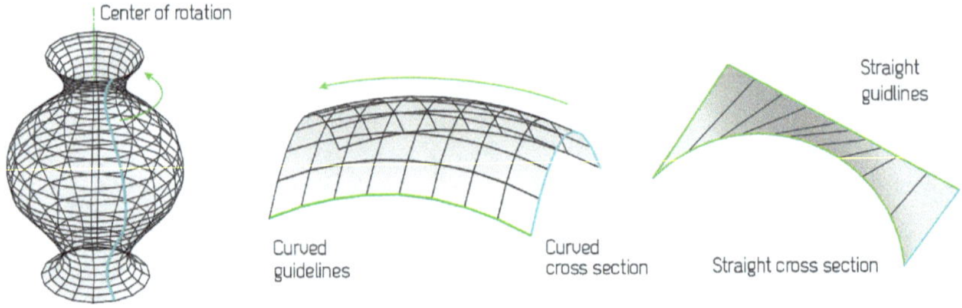

Figure 4.2 Rotation surface, Translation surface, and Ruled surface (Stahr, 2008).

Free-form curves are relevant to the generation of free-form surfaces. These curves can be classified into Bézier curves, B-Spline-curves, and NURBS-curves. Free-form curves are controlled through control points. These points are defined in design space, and their location and weight define the curve's form.

Bézier curves are parametrically modeled, polynormal curves that have two endpoints (anchor points), and control points (handles, tangent points) wherein two of them are attached to each anchor point. Bézier curves are a result of linear interpolation between two curves. For example, to construct a second degree Bézier curve, three control points P_0 , P_1, P_2 and the guidelines $\overline{P_0 P_1}$ and $\overline{P_1 P_2}$ are defined in the design space. The guidelines are parametrized, where t=0 is at the start points of the guidelines and t=1 is at the endpoints (Figure 4.3).

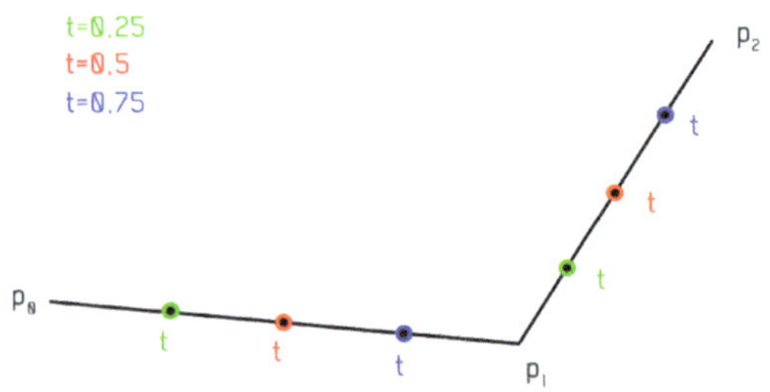

Figure 4.3 Division of the guidelines through different parameter t (Kondic, 2015).

The points with the same parameter are connected through a straight line. This line is then divided at the same parameter t, which represents the final points to generate the second grade Bézier curve (Figure 4.4) (Kondic, 2015).

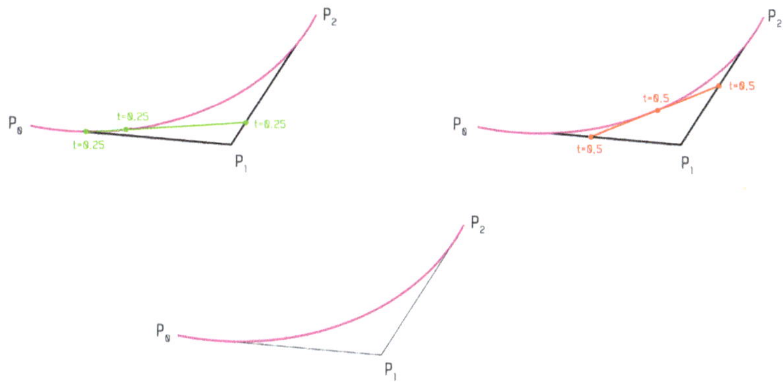

Figure 4.4 Bézier curve second grade (Kondic, 2015).

B-Spline curves consist of several Bézier curves that are connected at the endpoints. The endpoints represent the transition points between the curves; therefore, they must have the same curvature to share the same tangent. Bézier curves degree is defined based on the control parameter t, while B-Spline-curves change form based on their degree (Figure 4.5)(Helmut Pottmann, 2010; Kondic, 2015).

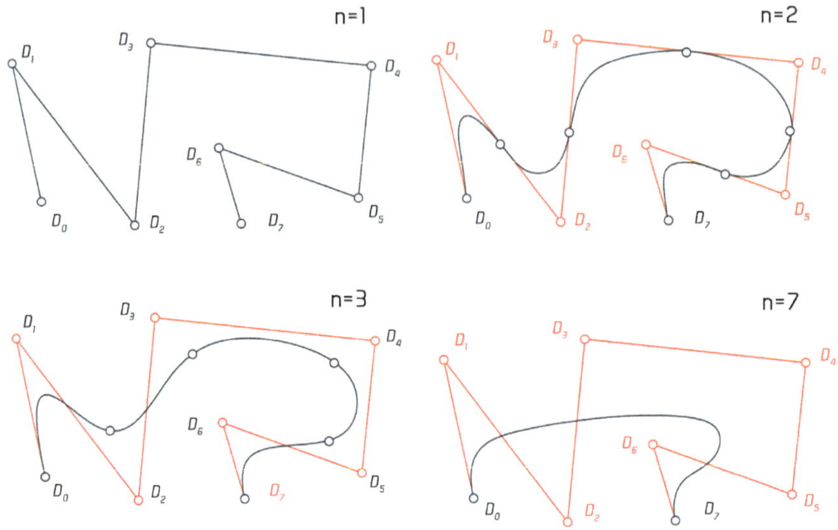

Figure 4.5 The effect of degree change on B-Spline-curves (Kondic, 2015).

Non-uniform rational B-Splines (NURBS) are similar to the B-spline curves that are controlled by control points. Also, NURBS's control points have a weight that influences the form of the curves (Figure 4.6). B-Splines are NURBS-curves with constant weight (Helmut Pottmann, 2010).

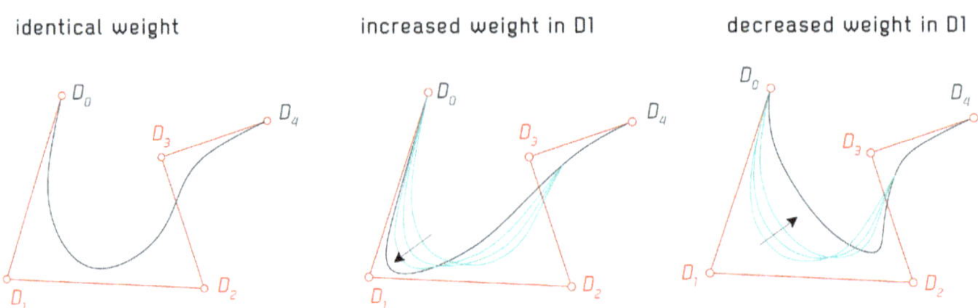

Figure 4.6 The effect of different weight on NURBS-curves (Kondic, 2015).

NURBS-curves allow the generation of any complex forms as a combination of small sub-areas/surfaces. NURBS method is the most commonly used in 3D modeling software nowadays. The difference between algebraic forms and NURBS-surfaces is that the algebraic forms are defined through fixed equations. However, NURBS-surfaces need a complex construct of mathematical logic that interacts in an iterative way to specify or create the requested form. Usually, the designer needs to define the boundary lines of the object, and then to divide the object into smaller sub-surfaces. A direct/continuous connection between the sub-surfaces has to be achieved to grant the generation of the surface. (Sören Stephan et al., 2004).

Free-form surfaces can be generated through various methods such as dilative translation, revolution, ruled, and NURBS-method (Figure 4.7&8).

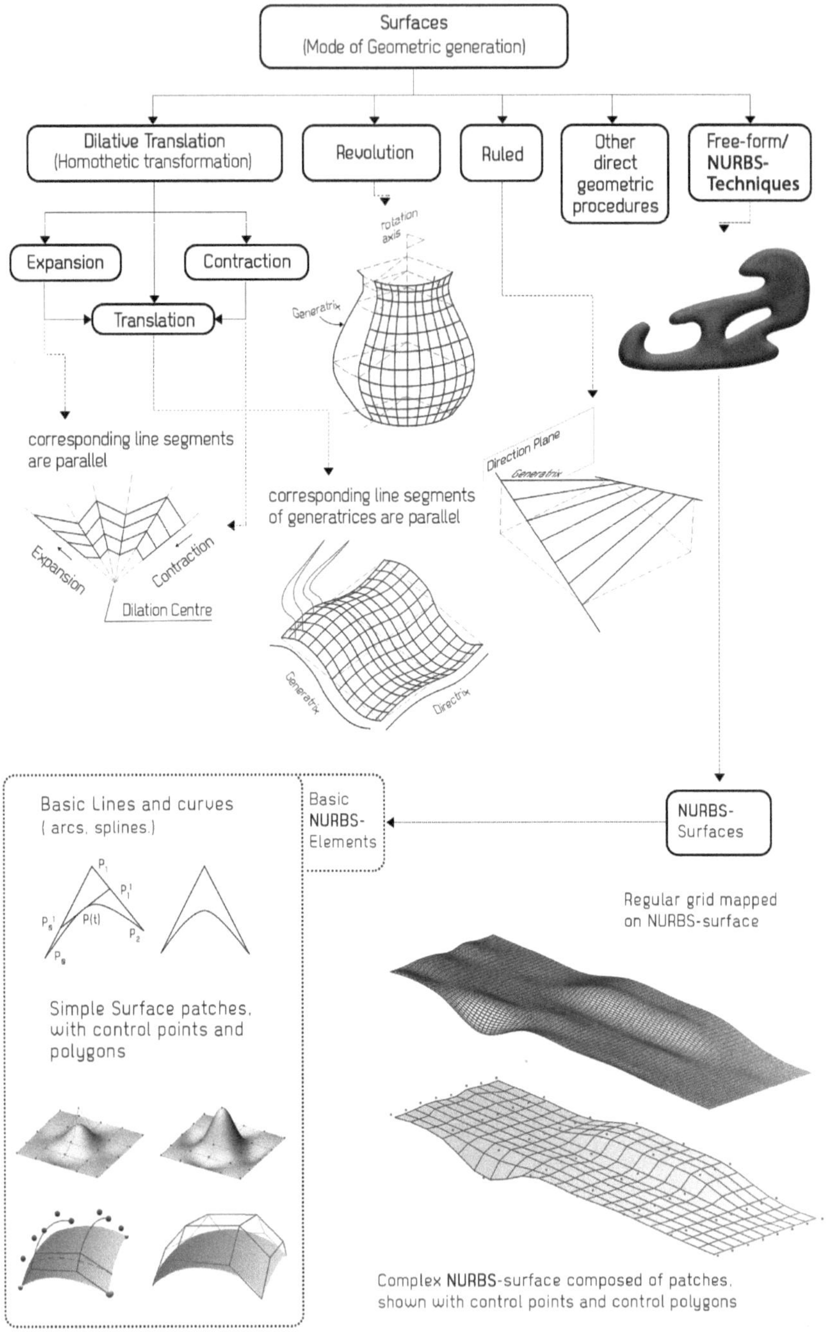

Figure 4.7 Different method to the geometric generation of free-form surfaces (Soeren Stephan et al., 2004).

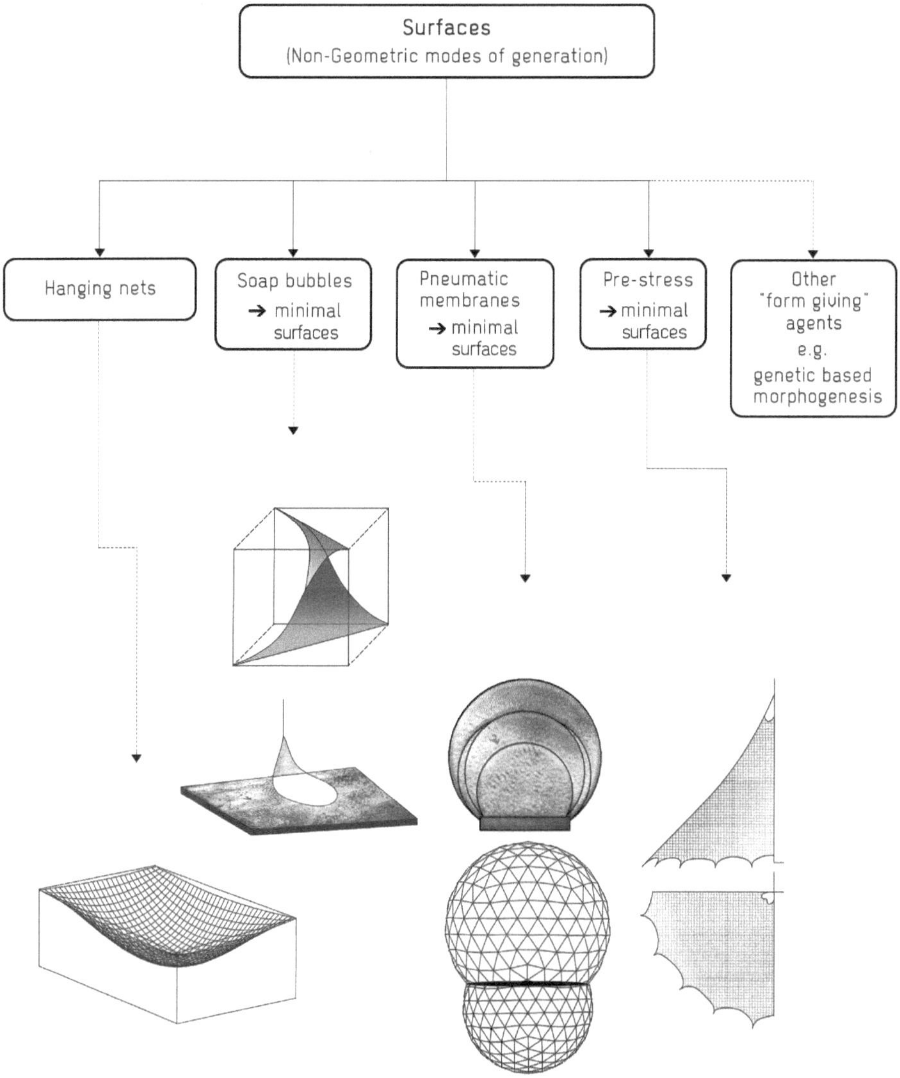

Figure 4.8 Different method to the non-Geometric generation of Free-form surfaces (Soeren Stephan et al., 2004).

4.1.2 Single-layered shell structures

The realization of free-form surfaces in façade/building industry is based on the tesselation and modeling of these free-form surfaces in manufacturable parts. Curved profiles and surfaces are usually not preferred because of the high manufacturing costs and the complexity of the planning. Therefore, the tendency is to transform the free-form surface into straight bars, planar surfaces, and connecting nodes. These elements are used to form a shell structure that is an approximation of the designed surface.

Surface meshing is the subdivision of irregular, curved, three-dimensional surfaces. The process defines the inner as well as the outer structure and appearance of free-form surfaces. Similarly, in the tessellation of a planar surface, the surface mesh consists of straight-one-dimensional-edge, zero-dimensional-node, and two-dimensional-partial-surface. These surfaces are joined at the common edges to form the mesh. There are many geometrical possibilities to generate meshes, but based on structural reasons, triangular or quad meshes are often used, and in some cases, irregular meshes are used that are generated based on NURBS-curves.

Zero-dimensional-nodes are signs for a position in space. This position is defined through coordinates. Straight-one-dimensional edges are a representation of a distance in space. This distance is defined by start and end points.

At each node in a mesh connects at least two edges. These edges are intersecting, arbitrarily aligned in space, and form a spatial (solid) angle. This angle is the adjacent edge angle $\varnothing_m$. It defines the base angle for the local node geometry, and its individuality is based on the mesh geometry and the meshing method (Figure 4.9).

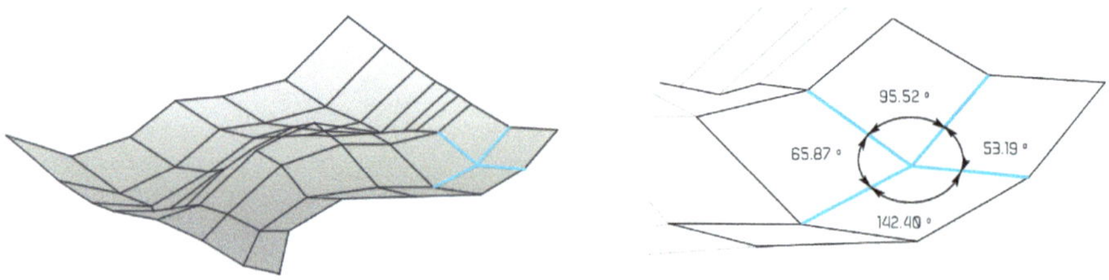

Figure 4.9 The adjacent edge angle (elliptically curved node) (Stahr, 2006).

The number of angles $\O_m$ at a node is identical to the number of ending edges at the same node. With respect to the sum of adjacent edge angles $\O_m$, the local mesh curvature can be classified in three statuses (Alber & Gabriel, 1983; Stahr, 2008):

1. $\O_m < 360°$ - elliptical curvature.
2. $\O_m = 360°$ - planar or two edges at a node.
3. $\O_m > 360°$ - hyperbolic curvature.

The further angle is identified between two adjacent planar surfaces. This angle is the adjacent mesh angle k_m, and it is strongly related to the structure and the curvature of the mesh (Figure 4.10).

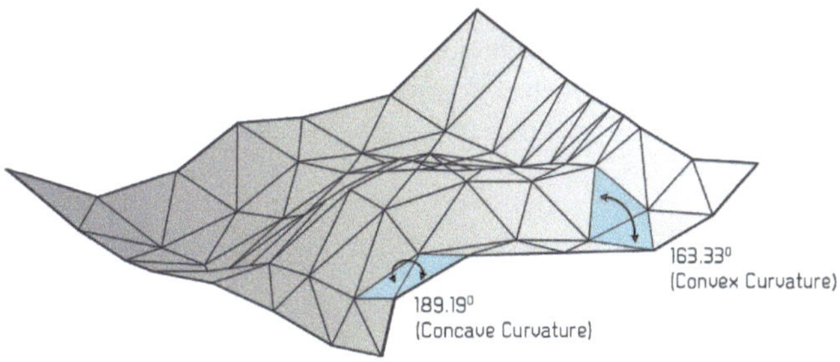

Figure 4.10 The adjacent mesh angle (Stahr, 2006).

The neighboring mesh angles k_m that are orthogonal to the net edge can be interpreted as the surface curvature. Again, three statuses can be classified (Alber & Gabriel, 1983; Stahr, 2008):

1. $k_m < 180°$ - convex curvature.
2. $k_m = 180°$ - planar.
3. $k_m > 180°$ - concave curvature.

In order to realize a free-form surface, it is logical to generate a single-layered shell structure that is coupled to the Free-form surface and consists of straight bars (profiles), planar surfaces, and connecting nodes. For the structural integrity of single-layered shell structures, the form of the planar partial surfaces, the cross-sections of the bars (profiles), the orientation of the profiles, and the selection of the nodes are relevant.

In shell structures, the longitudinal bars (profiles) axes are represented as straight lines that intersect in the connecting nodes. It is assumed that the bars have a constant cross-section. Therefore, the axes run through the center of gravity of the bars (profiles). The partial planar surfaces are enclosed by the axes of the bars (profiles) and are called facets. The linear contact between the bars (profiles) and the planar filling surfaces in shell structures act as linear support (Stahr, 2008; Sören Stephan et al., 2004).

For the orientation of the bars, it is logical to develop a local coordinate system at each node. If the surface is planar, the bar/profile axis is orthogonal and parallel to the surface plane. For a surface with constant curvatures – such as spherical surface – the local bars axes are aligned to the center point or the main axis. This approach could not be applied for free-form surfaces because of the irregular surface curvature. The bars/profiles axes in free-form surfaces are defined through local parameters, such as tangents and normals at the nodes. In general, there are two types of vectors for a shell structure. These are bar/profile vectors and node vectors. The node vector represents the local, normal, node axis, and it is the sum of all vectors that meet at this node – assuming planarity of the surfaces (Figure 4.11) (Soeren Stephan, Sánchez-Alvarez, & Knebel, 2004).

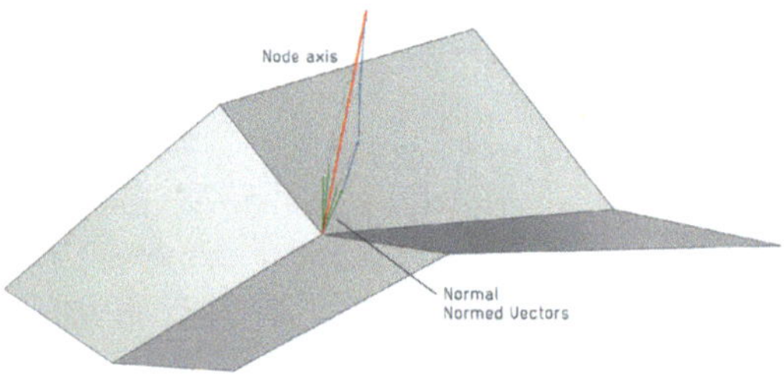

Figure 4.11 The formation of node vectors (Stahr, 2006).

The normal vector of the edges represents the bar/profile vector. These vectors are generated based on the incidence between an edge and its neighboring meshes. Assuming the planarity of the meshes, the neighbor mesh angle k_m is defined. Its bisector defines the normal vector of the edge. Based on the mesh edge and its normal vector, a local coordinate system can be defined through the third axis that is orthogonal to the two oriented vectors. The mesh edge is the x-axis, the edge normal is the z-axis, and their associate third axis is the y-axis (Figure 4.12) (Stahr, 2008; Soeren Stephan et al., 2004).

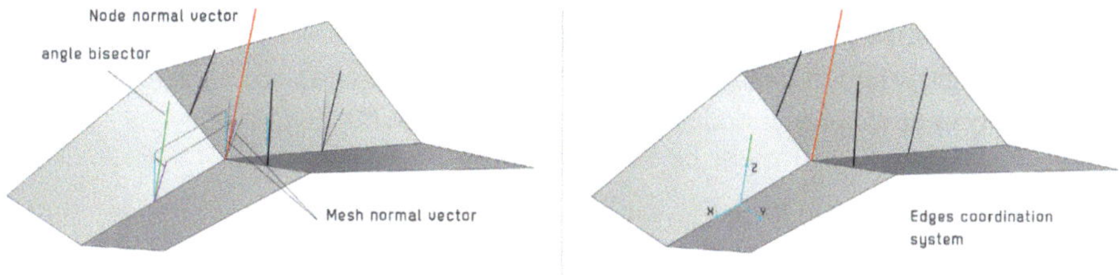

Figure 4.12 The determination of nodes normal vector and edges coordination system (Stahr, 2006).

The generated axes form linear constructive elements that form a geometric mesh. They are transformed into a constructive, structural surface mesh that is in both topological and geometrical relation. The node position in this mesh can be represented in the form of three polar angles: slope angle (vertical angle), sector angle (horizontal angle), and rotation angle (angle of twist) (Ruth & Stahr, 2005; Sören Stephan et al., 2004).

The relation between the node axis and the edge connected to it is an intersection between two lines, which will define an angle in all cases. This angle is called the slope angle αn,m. Based on the geometric individuality of shell structures, the angles between the normal vectors of the node and the edge axis are varied based on the local curvature of the mesh. Accordingly, certain geometric conditions can be defined (Figure 4.13) (Stahr, 2008):

$\alpha_{n,m} < 90°$ - concave elliptic curvature.
$\alpha_{n,m} = 90°$ - planar.
$\alpha_{n,m} > 90°$ - convex elliptic curvature.
$\alpha_{n,m} < 90°$ U $\alpha_{n,m} > 90°$ - hyperbolic curvature.

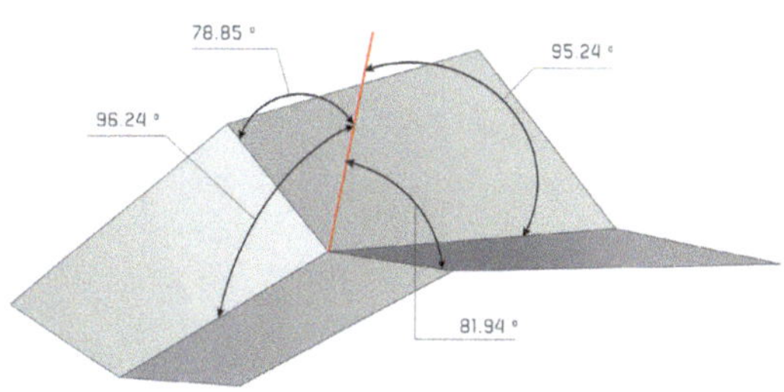

Figure 4.13 The slope angle (Stahr, 2006).

Through projecting the edges that are connected at one node onto the node plane, the three-dimensional orientation is transferred into a plane two-dimensional orientation. Accordingly, an angle can be defined between every two adjacent edges. This angle is called a sector angle $\beta_{n,m}$. Sector angles are not constant in a single mesh based on the variable geometric conditions. The sum of all sector angles at a node is 360° and a reflex angle can occur at the corner nodes (Figure 4.14) (Stahr, 2008).

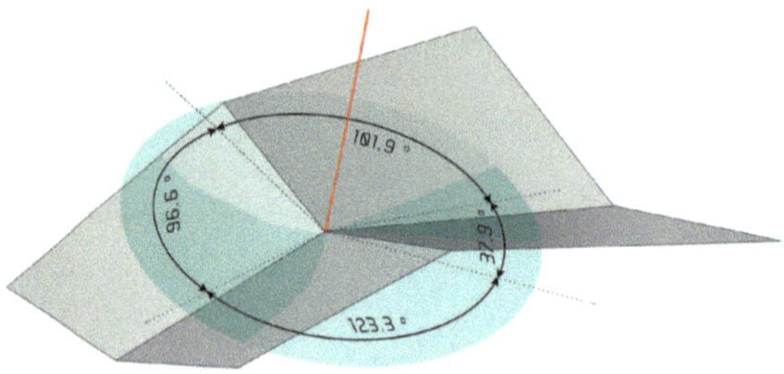

Figure 4.14 The sector angle (Stahr, 2006).

The normal vector of the mesh edge (z-axis) and y-axis of the local coordinate system, define a plane that is orthogonal to the mesh edge. The projection of the node axis on this plane allows the definition of anther polar angle. This angle is the rotation angle $\gamma_{n,m}$ (Figure 4.15) (Stahr, 2008).

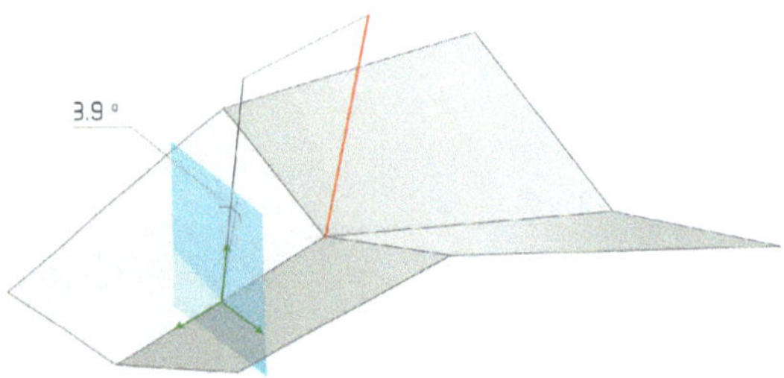

Figure 4.15 The rotation angle (Stahr, 2006).

Generally, the planning of single-layered mesh structures is a complex process because of various parameters, such as the length and cross-sections of profiles, the number of profiles connected to one node, and irregularity of free-form surfaces. Accordingly, inhomogeneous load cases are usually developed in such structures. Tensile, compressive, and bending moments are usually developed, which have to be accommodated by the nodes in addition to the change of the geometric parameters. Therefore, node connections are relevant to the whole frame structure (Sören Stephan et al., 2004).

Frame structures consist of main bars/profiles, node points, the connection between bars/profiles and nodes, and planar filling elements, with respect to the definition of the node (refer to 2.3.3). The bars/profiles with their straight longitudinal plane serve to form the structure, and to act as linear support for the planar filling elements. They are dimensioned based on the material properties and the structural model. The bar/profile element consists of the base bar/profile and end connection. The cross-section of the profile varies between rectangular, circular, hollow, or solid. It is crucial that the chosen profiles can transfer the present local load conditions without influencing the design aspects. The slenderness of the bars/profiles and the constant width grant a narrow supporting structure. It is preferred (if possible) to unify the height and width of the profiles.

The profile's orientation has to avoid any additional eccentricity-related loads. The strong axis of the profile has to be aligned in the direction of the sum of normal vectors of the adjacent meshes (normal vector of the edge). The bars/profiles act as linear support. Therefore, their cross-section has to provide specific stiffness and minimum width (Stahr, 2008).

The nodes differentiate based on their load-bearing and geometric capacity. In comparison to the bars/profiles, the nodes must guarantee local and global load-bearing capacity. Accordingly, the subjected loads transmitted through the bars/profiles have to be absorbed, as well as the tolerances in the structure that may occur because of manufacturing deviations.

There are two basic categories for the nodes. The first is the splice connection, where splice plates connect the node to the bars through a bolted or welded connection. These connections can be entirely or partially manufactured. Forces are transferred through the dimensioned bolted connections (defined by the number and arrangement of the bolt), or a dimensioned welded connection (defined by the length and the width of the weld joint). The second is the star connection, where the connection between the node and the bar/profile is perpendicular to the bar/profile axis. Forces are transferred through bolted head plate joint or welded connection, and the nodes are geometrically more flexible.

Different options of joining are used, such as the Butt connection that is used for star connections (Figure 4.16 a.b.), the Recess connection that is used for splice connections (Figure 4.16 c.d.). A decomposition of the node core can be applied (Figure 4.16 e), and a multi-layer structure can be utilized (Figure 4.16 f). The joining of the components should be done inconspicuously, collision-free, and without significant dimensional discontinuities (Sören Stephan et al., 2004).

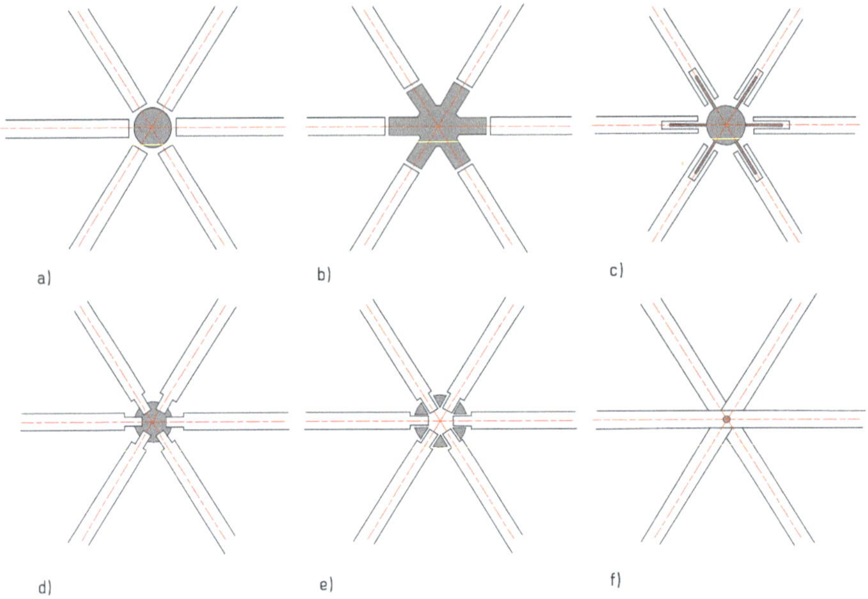

Figure 4.16 Geometry of different connecting nodes (Schätzke, 2016).

From an architectural point of view, it is preferred to unify the node type, bar/profile type, and connection detail to avoid irregular load distributions within the same structure.

The filling elements are often glass panels. The most economical forms are rectangular forms because of waste reduction. Any kind of other form increases the manufacturing costs. Based on how the glass material behaves, its crucial to use appropriate gaskets (EPDM for standard façade profiles) to avoid any damage to the glass. Distance blocks are usually used to control the movement of the glass panels within the bars/profiles (Ulrich Knaack, 2018).

For free-form surfaces, two glass panels are rarely planar to each other, and have the same loading condition. Therefore, the bars/profiles width is relevant to provide an adequate joint/support width between the glass panels.

4.2 Geometry generation (tessellation)

Free-form façades are complex systems. In order to generate a bar/profile network and to subdivide a free-form surface, the basic mathematical knowledge to tessellate a surface has to be present. Tessellation (tiling) is the creation of a complete, non-overlapping covering by one or two forms in a repetitive pattern. Tessellation in architecture goes back to the Sumer culture around 4000 years before Christ. They decorated their houses with differently colored clay bricks. In ancient times, small square blocks were used as decorative mosaic tiles that were called "tessella," which derives from the English word "tessellation". In Islamic architecture, tessellation was present in the form of significant use of mosaic patterns or ornaments. In 1619, German mathematician and physicist Johannes Kepler presented "Harmonica mundi," where he showed a tessellation concept based on his research on the hexagonal honeycombs and snowflakes. He could tessellate a plane without gaps, overlaps, and periodic tilting. In 1890, Russian mathematician Kristallpgraph Jewgraf S.Fedorow proved that there are only 17 different methods to tessellate a plane periodically (Heesch & Kienzle, 1963; Scriba & Schreiber, 2010).

In 1982, the mathematician Roger Penrose presented for the first time the aperiodic (quasiperiodic) tessellation theory. He proved that it is possible to divide a surface without a periodic repetition. Thus, the crystals and their arrangement had to be redefined. Tessellation is even used in the industry. A method has been developed to come up with a low waste method to divide cement plates into congruent parts. Every now and then, someone is coming up with a new idea of tessellation that changes its basic principles (Oberlerchner, 2015).

There are many possibilities to tessellate a plane. Accordingly, the different basic tessellation methods and their parameters have to be clarified. Tessellation is defined as the division of a plane into a complete pattern with overlap-free polygons that cover the entire plane (Frettlöh, 2002; Oberlerchner, 2015).

The polygon is an elementary geometry with n-corners ($n \geq 3$) that forms a closed line/path, which passes through n different points on a plane. Polygons can be classified into convex polygons, where all inner angels are smaller than 180°, concave polygons where at least one angle is larger than 180°, self-intersecting polygons, where two edges are not only intersecting at the corner points, and not self-intersecting polygons, where a corner point lies on one edge of the polygon (Wittmann, 1987).

Tessellation has many different types: monohedral, where only one form type is used, the tiles share a corner or an edge, and a corner never lies on edge (edge-to-edge); isohedral, where a monohedral transferred through mirroring, rotation, or moving to form the tessellation (tiling); k-hedral, where the tiling is done through k number of tiles to form the final tessellation; and k-morph (monomorphic), where the tiling cell allows different types of arrangements to form different tessellation arts (Oberlerchner, 2015).

Tessellation can be divided into periodic and aperiodic styles. Periodic tessellation can be as platonic and Archimedean. Aperiodic (quasiperiodic) tessellation contains local periodicity but is not globally periodic, such as Penrose tessellation.

There are various possible ways to tesselate a freeform surface. There are many examples of two-dimensional and three-dimensional tessellations in architecture. Some of the commonly applied tessellations in architecture are described in the upcoming section.

4.2.1 Periodic Tessellation

Periodic tessellation is a transformation in two different independent directions as a congruence mapping. The two different independent directions are two linearly independent translation vectors that create a regular grid. The transformations are processes that lead the tiling process to form congruence mapping or symmetry conditions. The tile does not change form based on these transformations, as the translations are invariant, which means that the distance/angles between two points within the tiles are constant. These transformations are in the form of translation (moving), rotation, mirroring, or glide reflection (transflection) (Oberlerchner, 2015).

As mentioned before, the mathematician Jewgraf S. Fedorow classified the symmetry options into 17 symmetry groups (Crystallographic groups). These groups can accurately identify periodicity. However, only rotations of 180°, 120°, 90°, and 60° were permitted (Heesch & Kienzle, 1963).

Some periodic tessellation can be present in Platonic (regular) tessellation, Archimedean (Semiregular) tessellation, Demiregular tessellation, and irregular tessellation.

Platonic (regular) tessellation is when the corner points of a polygon meet at node points and the node points are congruent to each other. Platonic tessellation has to contain only regular polygons of the same type, to apply the edge-to-edge rule, to result in 360° for the sum of the inner angles that meet at a node point, and to have the same number of polygons that meet at a node point. The regular polygon is defined as equal-angled and equal-sided, so that its corners lie concyclically on a common circle. Both share the same center point. When the corner points of the polygon connect to the center point, n congruent isosceles triangles can be formed. As a result, trigonometry could be applied to collect information concerning the side length and the angles of the polygon (Heinrich, 2014).

Equilateral triangles (3,3,3,3,3,3), squares (4,4,4,4), and hexagons (6,6,6) are suitable for platonic tessellation. They are expressed by how many n-corners meet at a node point, and it is usual to count the polygons counterclockwise. In 1693, Johannes Kepler defined three types of platonic tessellation (Figure 4.17) (Grunbaum & Shephard, 1987).

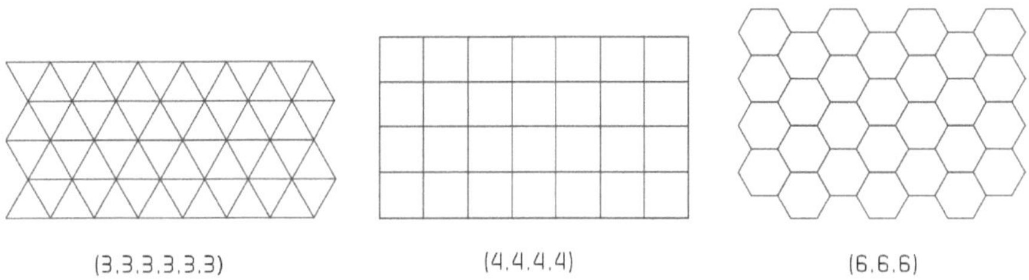

Figure 4.17 Platonic tessellation.

Archimedean (Semiregular) tessellation is when several regular polygon types are used, and they have congruent node points. Based on the angle rule from Platonic tessellation, only three types of polygons can be used. Accordingly, it is essential to define the different cases when and which two or three polygon types are able to meet. In the case of three, four, or five polygons meeting at a node point the corners of the polygons have to be all in an even or odd number. If one of the polygons has an even number of corners and the other two polygons have an odd number of corners, the congruence of the nodes cannot be achieved. There were eight variants of the Archimedean tessellation that could be defined; three polygons (4,8,8), (4,6,12), and (3,12,12), four polygons (3,6,3,6), and (3,4,6,4), five polygons (3,3,3,3,6), (3,3,3,4,4), and (3,3,4,3,4) (Figure 4.18) (Grunbaum & Shephard, 1987; Heinrich, 2014).

Demiregular tessellation is similar, as regular polygons have to be used and the edge-to-edge rule still applies. However, the congruence of the nodes is not achieved. Accordingly, there are two different types of nodes that are in the system. 20 types of Demiregular tessellation have been defined.

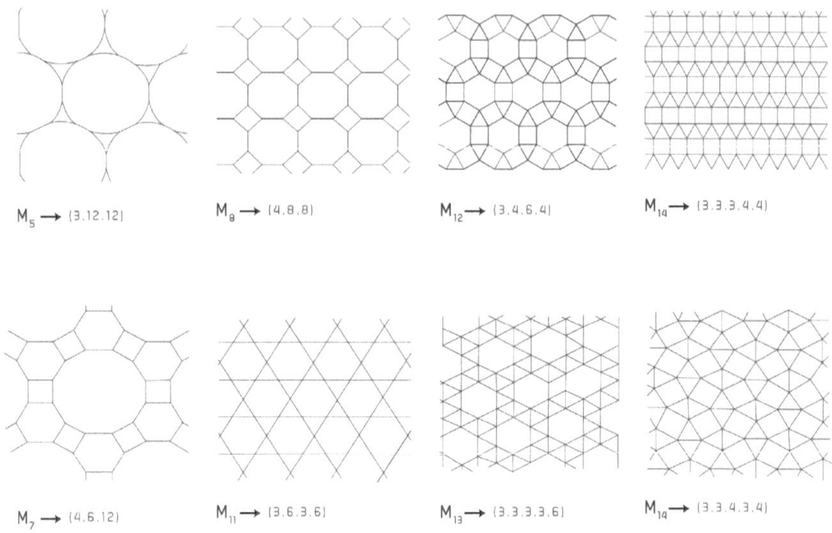

Figure 4.18 Archimedean tessellation.

Irregular tessellation is created when irregular polygons do not have the same inner angles and the same sides. However, it still forms a monohedral tiling, which means that the tiles are congruent but not the node points. In 1918, mathematician Karl Reinhardt proved that monohedral tiling is only possible with a number of corners $n \leq 6$. Parallelograms, rectangles, and any irregular quad can be arranged with no gaps. Convex pentagons are interesting as not all possible combinations are known. Fifteen recognized types of convex pentagons could allow tessellation. It applies that two inner angles of the polygon must result in 180° (Figure 4.19) (Reinhardt, 1918).

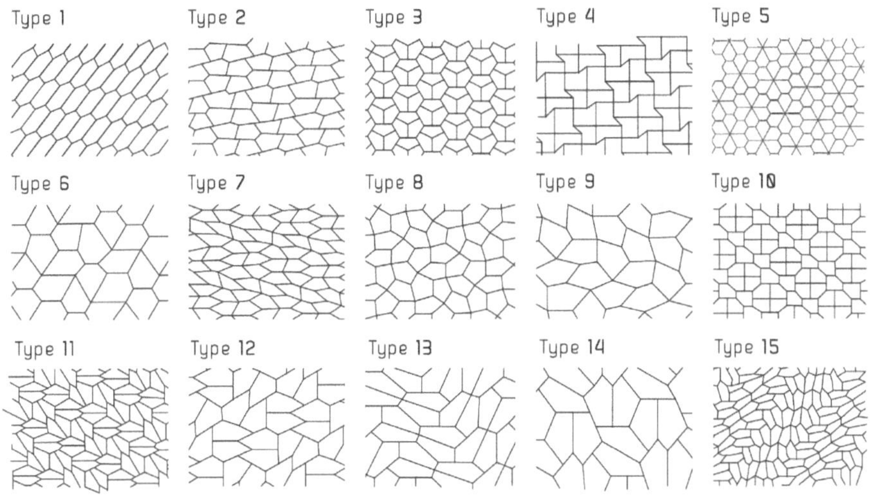

Figure 4.19 Monohedral tessellation with pentagons.

Later, K. Reinhardt proved that there are three types of hexagons, where two neighbor angles do not sum 180°, that can form monohedral tessellation (Figure 4.20) (Reinhardt, 1918).

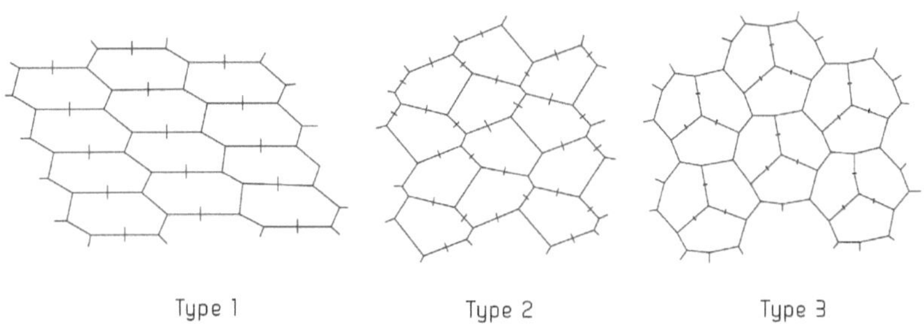

Figure 4.20 Monohedral tessellation with pentagons, Karl Reinhardt.

4.2.2 Aperiodic Tessellation

Aperiodic tessellation is when the polygons are neither congruent to each other nor to the node points. It is not periodic. Therefore, it has no translation symmetry, which provides many possibilities of not periodic tessellations. Types of aperiodic tessellations are Self-similar tessellation and Penrose tessellation.

Self-similar tessellation is when the tiles can be arranged together to create an enlarged version of itself. This process is called inflation. Deflation is when the tile can be divided into several smaller replicants. One example is Sphinx, where through deflation the module can be deconstructed in smaller congruent tiles. The arrangement is not periodic. Therefore, the modules can be moved or rotated (Figure 4.21) (Gardner, 1997).

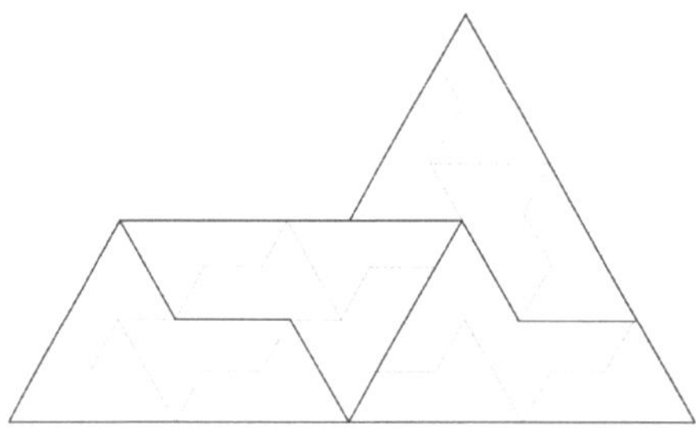

Figure 4.21 Aperiodic tessellation of Sphinx.

Penrose tessellation is when the tiling is neither periodic nor symmetrical. However, periodicity occurs at smaller local regions. The tessellation order is strict, and the periodic areas can be infinitely represented. In 1974, British physicist Roger Penrose developed several tiling options that consisted of two different types of tiles. He proved that it is possible to cover a plane with an infinite number of combinations with no gaps, using two different geometrical forms without repeating larger areas.

In 1977, Penrose published his results in "Pentaplexity – A Class of Non-Periodic Tiling of the Plane". He used pentagons, five-pointed (stars), diamonds, and boats to form his tessellations. The shapes have to follow specific matching rules to ensure no gap tiling.

Penrose tiles are formed through the subdivision of pentagons into smaller pentagons, which results in triangular gaps (A). When the deflation is repeated, gaps occur in different forms (B). In the third stage of subdivision, the diamond form is developed (C) that can be divided into a star and boat forms (D) (Figure 4.22) (Penrose, 1979).

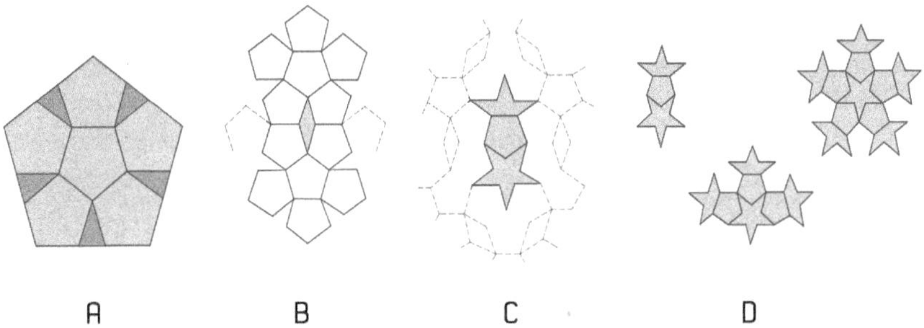

Figure 4.22 Different stages of subdivision to generate Penrose tessellations.

Penrose reduced the tiles into narrow and thick diamonds, and darts and kites. The matching rules present certain rules of arrangement/combination between the tiles. For example, the darts have to be filled with two kites. The inflation and deflation methods can be applied, which allows the formation of aperiodic tessellation. The tiles can be repeated infinitely, but when located in a different environment, the aperiodicity of the pane remains (Figure 2.23) (Penrose, 1979).

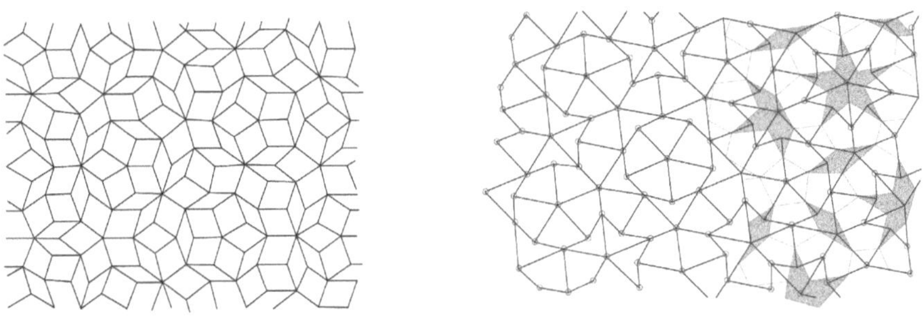

Figure 4.23 Penrose tessellation through Diamonds (left), Darts and kites (right).

4.3 Façade functions

Façades are important as they are the first element of the building that an occupant interacts with. These interactions are both visual and functional. The functional interactions are based on what the façade can provide to the occupant. The occupant may consider these functions as beneficial as well as detrimental. Ulrich Knaack, in his book "Façades" offered this description (Figure 4.24):

"Façades are not limited to the actual space they occupy as part of the entire structure, but also influence the space in and around the building. A façade is the key element when observing a building from the exterior and has impact on the interior. View, lighting, ventilation, user comfort, some building services and possibly loadbearing are all tasks the façade may need to address."(Knaack, Klein, Bilow, & Auer, 2014).

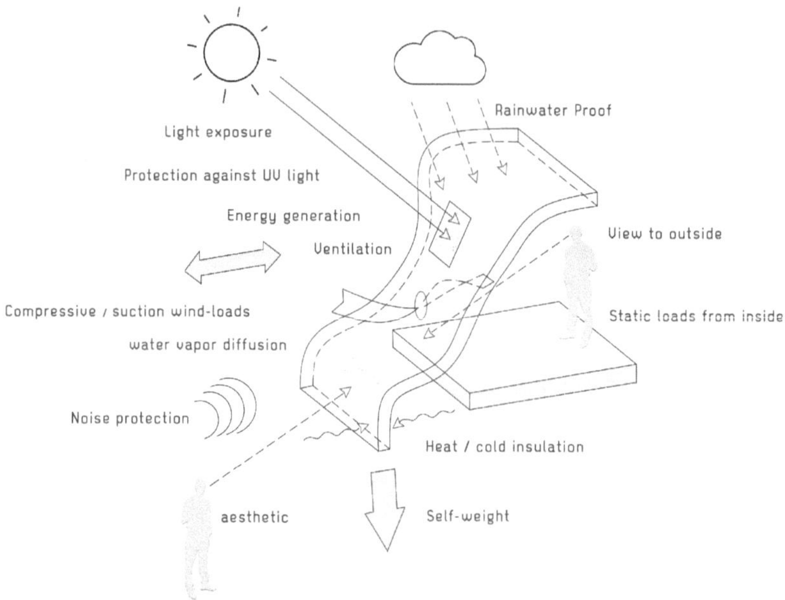

Figure 4.24 Façade external influences (Knaack et al., 2014).

The decision-making phase at façade design and construction is controlled through the required functions that the façades need to fulfill. Façade functions are developing/changing through time. Standard main façade functions are: insulation against heat, cold and noise; waterproofing; providing natural light for the interior; protection against light; resistance against wind loads, interior loads and self-weight; vapor management; visual contact to outside and inside; and architectural appearance (Klein, 2013; Knaack et al., 2014).

Different classifications had been done to investigate façade functions, e.g., façade function tree p.110 (Klein, 2013). To simplify the classification, façade functions can be subdivided into three categories: primary functions (protection of outer influences), secondary functions (user-comfort and economics), and subordinate functions (aesthetics and structural integrity). A combination of these functions based on the requirements must be aligned with each other in the right ratios to ensure a successful façade (Beyer & Kuba, 2006).

The primary function of façades is the protection against external influences. The first façade ever built was designed solely for this purpose. It is considered beneficial to the occupant.

The first people went into caves to seek protection against the cold. Cold protection is the main primary function of a façade. The façade must be sufficiently insulated so that the energy gains are always higher than energy losses, which is the case in winter. The problem is that in summer, the exact opposite happens. Accordingly, solar absorption should be kept to a minimum by means of choosing suitable materials and systems. In order to protect the occupants and the building itself, humidity should be kept outside. Humidity attacks the building either through the rains or air humidity. Façades have to contain a suitable drainage system that ensures the flow of rainwater to the outside of the building. Air humidity enters the building through diffusion. If the façade is incorrectly constructed, condensation can be formed inside the façade. This water will influence both façade performance and its life cycle. Air drafts have to be as low as possible, thus the façade construction has to be airtight, which supports the thermal performance of the façade through reducing thermal loses (Richter et al., 2008; Ulrich Knaack, 2018).

The main two aspects of noise protection are sound insulation and connection types. Sound insulation is concerned with preventing noise caused by external environmental influences such as airplanes, cars, animals, or people from penetrating into interior spaces. This happens through the transmission of vibrations via the air (airborne noise). Massive façades are effective at blocking sound vibrations. Besides, there are different kinds of materials that offer good sound insulation. The performance of these materials is based on layer depth. Connection types are concerned with the prevention of spreading the inside noise through the façade. It is essential to design adequate separation within the connecting points to prevent the spreading of impact noise between interior spaces (Richter et al., 2008; Ulrich Knaack, 2018).

In case of fire, a sufficient amount of time has to be granted to the occupants to be able to evacuate the building. Therefore, fire protection has to be considered while designing the building/façade. Building materials are categorized into not flammable and flammable, or high fire-resistance, normal fire-resistance, and low fire-resistance. Low fire-resistance materials are not usually preferred. It is essential to choose the right material based on the classification of the building. Buildings have fire-resistance classifications that identify the amount of time for emergency evacuation that has to be granted to the occupants, and which kind of materials are suitable for each classification.

Exhaust gases and dust are released daily into the air from both industry and transportation. Specific components of these exhaust gases and dust may interact with the façade structures, which may lead to several consequences such as optical defects, change of colors, deterioration of the insulation properties, and degradation of structural integrity. Accordingly, the façades grant protection from pollutants to the occupant (Beyer & Kuba, 2006).

Depending on the material, façades can deteriorate through weathering in different ways. Wood, for example, has to have special treatments to enhance its ability to withstand wood-destroying insects and rot caused by moisture. There are different kinds of coating that have to be applied to protect wood against moisture. Furthermore, it is convincing to create wooden constructions as ventilated systems, which helps to evaporate the moisture more easily. Similar to the rot that occurs in wood, corrosion affects metals. Moisture can cause mold growth in almost any material. Therefore, it is crucial to pay attention to the different materials that are used in construction, and to prevent any occurrence of damage like mold growth that can cause health problems. Thus, the façades help to provide health protection.

Finally, one occupant may be an external influence that requires protection from another. Therefore, the façades have to provide privacy and protection against burglary.

The secondary functions of façades are user-comfort and economics. The façade may be able to enhance the experience of the occupant in the building, within economical range if possible and/or required.

To achieve user-comfort, the façade has to grant constant temperatures inside the interior spaces all year. Thermal comfort is achieved when the heat balance of the body core is at 37° (equilibrium state). Thermal comfort cannot be calculated precisely. The process depends on the individuals inside the space. A pleasant indoor climate is defined as an environment where 80% of the occupants find it thermally acceptable. Other factors that influence thermal-comfort are clothing, degree of activity, air movement, relative humidity, and ambient air temperature (Willems, Dinter, & Schild, 2006).

The façade cannot control all of these factors. Clothing and activity level play a huge role. However, to guarantee a comfortable state for as many occupants as possible, the air exchange rate is a major factor. It is important to provide interior spaces with fresh air at sufficient intervals. The same applies to humidity (Richter et al., 2008).

Quality of light contributes to user-comfort. Natural light is always preferred to artificial light. However, glass elements and windows have to be carefully oriented to avoid glare. This should allow as much natural glare-free light as possible to enter interior spaces, but also not enough to cause overheating. Accordingly, all these factors have to be considered through the planning phases (Richter et al., 2008).

The costs that are involved in the manufacturing of façade, have to be in line with the purpose of the building to be considered economical. The functions are weighted differently depending on the client. Nevertheless, saving manufacturing costs is always beneficial to the economic efficiency of the façade. Operating costs can be lowered by using adequate solutions, which will improve the energy consumption of the building. Maintenance costs can be optimized by considering available parts in the market. Overall, for the façade to achieve appropriate market value, it has to match or enhance the use of the building and have the ability to adapt to new uses (Beyer & Kuba, 2006).

The subordinate functions are aesthetics and structural integrity. The façade is the center of building's appearance. It is used to contribute to the value image of corporations as well as private individuals. A well-designed, elaborate, or innovative façade attracts attention. The façade of a building represents its owner and can be conceived as part of the identity of the owner. Similarly, the occupants who are living or working in an iconic building are identified with it. Accordingly, complex buildings/façades are becoming more popular, and this presents a new challenge, which is structural integrity. Such structures are unknown fields and unique experiences. Most of the time, a customized solution is needed. Therefore, the structural integrity of such new structures has to be carefully designed and tested (Beyer & Kuba, 2006).

4.4 Parametric structural and geometrical optimization methods

Optimization, referred to as Mathematical optimization or mathematical programming, is the selection of the optimal element within a set of variables with respect to some criterion. The basic structure for an optimization model consists of three components. First, the variable set that defines the unknown quantities in the problem. These values lead the search engine to an optimal solution. Second, the variable domain that defines the allowable values that the variables can utilize to express a specific relationship between them. These relationships are referred to as constraints. In any design problem, these constraints define how the system performs to restrict or quantify the solutions. Finally, the objective function that describes the desirable properties that the model should improve. The optimization goals of the problem are encoded in the objective function (Nowatzki et al., 2013).

Many types of research have been done on the structural and geometrical optimization of truss structures, and on the cross-section optimization of single-layered shell structures. However, the focus is often on the form-finding and the geometrical optimization of triangular and rectangular structures, by means of the translation of their nodes and the orientation of their structural elements (Billington, 2003).

The design of single-layered shell structures over a given free-form surface has to meet structural, optical, manufacturing conditions. Often the engineers rely on their experience and intuition, which limits the number of possibilities that can be used. The goal is to find a structural approach that can fulfill the structural conditions like permissible stress or stability with a minimal amount of material. The boundary condition (in this case, the given free-form surface) defines the search range to find the best structural performance with minimum costs.

The stiffness condition of the nodes in shell structures makes the design process more complex. The nodes are not only subject to axial forces but also to moments. Because of the three-dimensional orientation, three moments will occur. Looking at the structure as a whole system with thousands of elements, the interdependency becomes incredibly complicated. The number of possible combinations and variables is gigantic.

There are two different design approaches to solve these structures: "Bottom-up," and "Top-down." The top-down approach is defining the overall concept, generating the final form, meshing the surface, and generating the required details. The bottom-up approach is defining the primary elements, meshing the elements, transforming the primary elements, and generating the final form (Figure 4.25) (Büscher, Christensen, Hansen, Mogensen, & Shapiro, 2009).

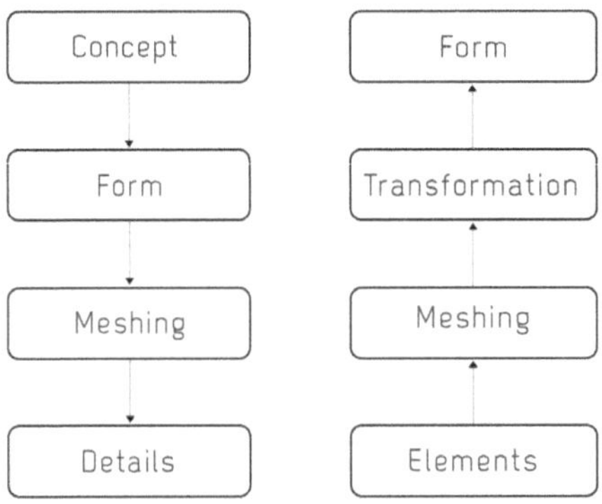

Figure 4.25 Top-down design approach (left), and Bottom-up design approach (right).

The top-down process starts with the overall concept, which is broken down into subsystems and separate tasks. This process allows larger teams to work on the parts of the project, and its subsystems can be easily combined.

The bottom-up process starts with the individual elements, which are assembled to form the complete system. A simple example of this process is LEGO design.

The two approaches can be applied in architecture and structural design. The generation of the single-layered shell structure can be granted, using both approaches (Figure 4.26). In the top-down approach, the shape is freely determined during the design process. The resulting geometry is then subdivided (tessellated) into individual cells. The geometry now can be optimized to find the best geometrical and structural conditions. Finally, the detailed design for the individual elements can be carried out.

Form-finding is a typical bottom-up approach. The design starts with an initial geometry and predefined topology of the elements. Under the determined load conditions, the final geometry is formed through the predetermined elements.

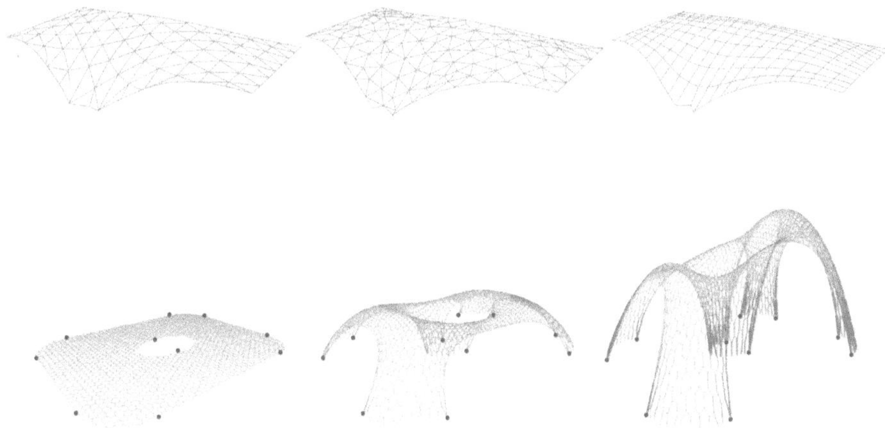

Figure 4.26 Example of the top-down design approach (top), and the bottom-up design approach (bottom).

The parametric modeling methods allow the generation of such geometrical data sets that can be discretely optimized. These data sets can describe and modify these geometries using CAD software through a finite set of parameters. The goal is to find the best-optimized shape for our structure within specific parameters. This addresses difficulties in two aspects:

- A vast number of variables: the controlled parameters of the model vary from a discrete number of elements, the orientation of the mesh, coordinates of the nodes, and the dimensions of the cross-sections of the elements.
- Complex objective functions: the optimization objectives for structural design deal with structural behavior of the elements by means of its deformation, material utilization, and the applied forces. The relationship between the input parameters and the results can hardly be described through mathematical functions or equations. Therefore, finite-elements methods are usually used.

Discrete optimization is a branch of optimization in applied mathematics and computer science. Like linear optimization, it deals with the optimization of linear objective functions over a set that is constrained by linear equations and inequalities. The difference is that in discrete optimization, some or all variables are limited to discrete variables. These variables can be assumed as a set of values such as integers, or result value.

Discrete optimization is about constructing a subset of discrete elements (nodes and elements) that correspond to certain constraints, and which is optimal with respect to an objective function (minimal weight, maximum stiffness). In such problems, only one optimal solution is not feasible. A set of feasible solutions is generated to reduce the number of solutions discretely. Accordingly, the best solution can be found (Cook, Cunningham, Pulleyblank, & Schrijver, 2009).

To solve discrete optimization problems, on the one hand, there are exact solvers such as Branch and Bound method and cutting-plane method, where the best solution yields the same results each time. On the other hand, there are heuristics solvers, where an approximate can be found, and the resulting output can change at each iteration. Several solution methods are usually combined to help solve discrete optimization problems, as customized algorithms must often be developed to find the best solution (Suhl & Mellouli, 2009; Tedeschi, 2014).

The most common algorithms that are used to solve discrete optimization problems are evolutionary algorithm, and multi-objective optimization. Two elements composed model can describe the optimization problems; the first element is the objective function, which will try to yield a minimum or maximum results/value; the second element is a set of variables defined by a range that influences the objective function (Tedeschi, 2014).

The search technique can be categorized into calculus base techniques, guided random search techniques, and enumerative techniques. Evolutionary algorithms lay within the guided random search techniques group (Figure 4.27).

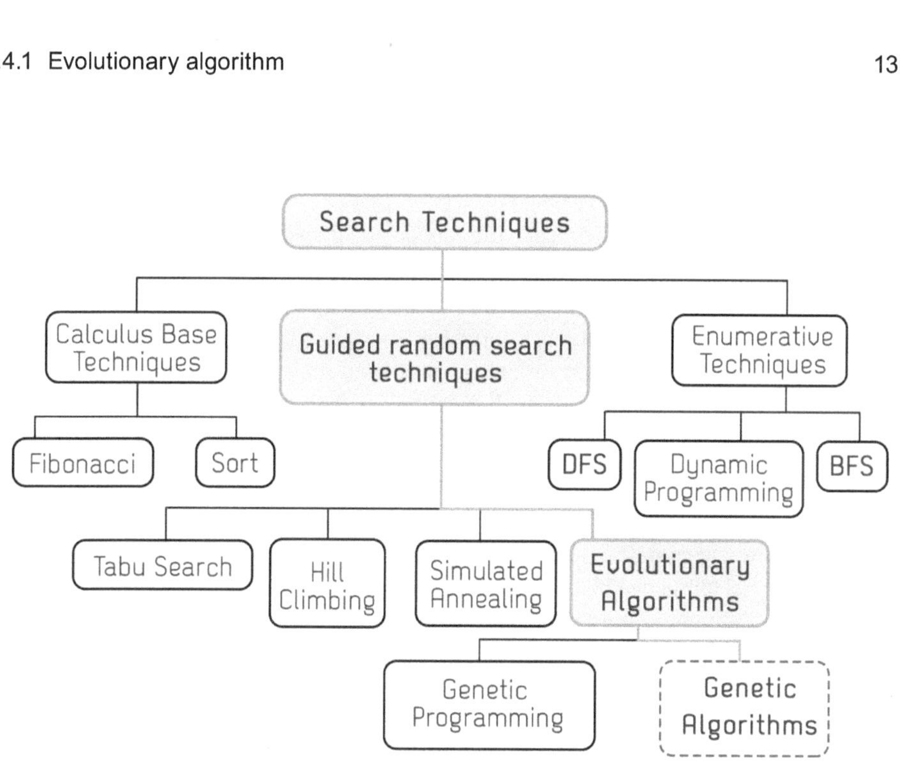

Figure 4.27 Current method for Optimal-search.

4.4.1 Evolutionary algorithm

Evolution is an art of optimization. Optimization does not yield perfection, whereas evolution allows finding a precise functional solution to certain problems that are caused by the organism's environment. The natural mechanisms evolved elaborately from the engineering perspective. The function is the only quality that is based on natural selection, and functionality is the result based on the optimization by iterative selection and mutation. Evolutionary algorithms are used often to solve complex engineering optimization problems. The conventional algorithm techniques such as gradient descent, purely random search, and deterministic hill climbing are not satisfying when it comes to nonlinear optimization problems that are stochastic, temporal, or contain chaotic components. Evolution allows computing solutions to a problem that appears to be intractable (Bäck, Fogel, & Michalewicz, 1997).

Evolutionary algorithms (EA) are a type of stochastic, heuristic optimization techniques inspired by the evolution of natural organisms. Following the rules of nature, possible solutions for particular problems can be artificially generated. The basic principles for EA are derived from nature and its selection methods. Charles Darwin described in his book "The Origin of The Species" the principles of natural selection, and therefore defined the basic rules of evolution.

Since the creation of the first cell, life has been achieved by reproduction (multiplication). Through this process, nature introduces small random mutations. These mutations introduce diversity, and in combination with the environment, unique individuals have been created. Each generation is alive for a certain period and some family groups reproduce more than others. Family groups that lived longest and reproduced the most passed on genetic information uniquely adapted to their environments, thus enabling their survival. The British philosopher Herbert Spencer used Darwin's idea, coining the term "survival of the fittest," where the best designs survive, and the bad ones disappear. This is the basic idea behind the optimization with evolutionary algorithms (Fogel, 2006).

The first references in the field of computation data go back to the early 1960s when Lawrence J. Fogel published his dissertation "On the Organization of the Intellect," which is considered to be one of the first dissertations in the field of evolutionary computation. In his 1986 book "The Blind Watchmaker," Richard Dawkins presented a small program that is based on human selection, where the generated evolutionary algorithm is constructed through the collective learning processes that may occur within a population of individuals (Bäck et al., 1997; Bäck & Schwefel, 1993).

Evolutionary Algorithms (EA) are mainly used to find optima. The basic logic is not surprising when we think about it as a fundamental problem that our head needs to solve. We follow the same logic every time we look for an acceptable solution for a fundamental problem. We tend to: (1) develop a set of possible, or random solutions, (2) evaluate and select some of them, (3) make some changes to them, (4) combine solutions to develop a new better solution, and finally (5) repeat steps 2 to 4 till we find a final satisfactory solution. This way of thinking can describe how an evolutionary algorithm works. The EA procedures can be formulated as (1) generation, (2) selection, (3) mutation, (4) crossover, and (5) iteration.

All evolutionary algorithms share basic instances. Initialization, where the EA establishes a collective learning process of a population of individuals and first generation of solution candidates is generated; Evaluation, where the individuals are evaluated based on a measure of quality or fitness factor and a value of the fitness function is assigned to each solution candidate based on its quality; and Iteration, where certain steps are followed until certain criterion is met, which includes the steps of selection, recombination (crossover), mutation, evaluation, and final selection (Figure 4.28) (Bäck et al., 1997).

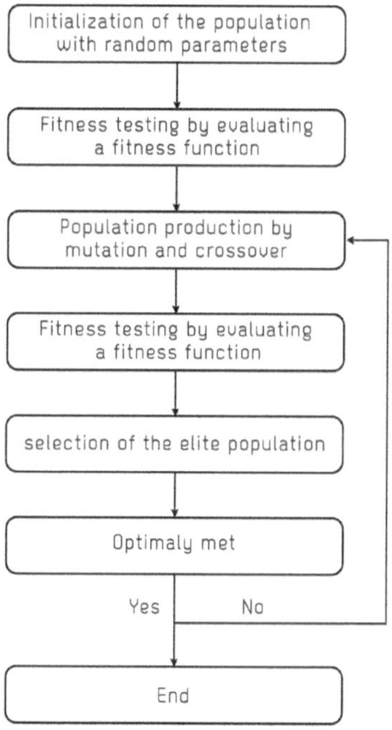

Figure 4.28 The sequence of Evolutionary Algorithms (EA).

Evolutionary Algorithms have only recently been introduced to the building industry, and there are not too many examples brought into existence. Nowadays, parametric designs are gaining interest because of the implementation of digital tools in architecture. EA are implemented in many software applications that allow the optimization of designs at the early stages of the planning processes (Fogel, 2006).

4.4.2 Multi-objective Optimization (Pareto optimization)

For many optimization tasks, it is possible to define several objectives that are fundamentally independent of each other, such as structural design, elements deformation, material utilization, and material costs. It is often not possible to optimize all objectives together, as the objectives usually conflict with each other. For example, when trying to increase the stiffness of the structure, this is only achievable with larger cross-sections.

Multi-objective optimization (Pareto optimization) is named after the Italian engineer and economist Vilfredo Pareto. He used it in his studies of economic efficiency and income distribution (Ehrgott, 2005).

A Pareto optimum is a state in which it is impossible to improve one objective without deteriorating the other objective. For a given system, the set of solutions that are Pareto-optimized are called the Pareto Front. It is the line to be considered the limit between feasible solutions and impractical ones. All possible solutions on the Pareto Front can be accepted as optimized solutions. Next step is the comparison, evaluation, and decision making of these solutions on different defined problems (Figure 4.29).

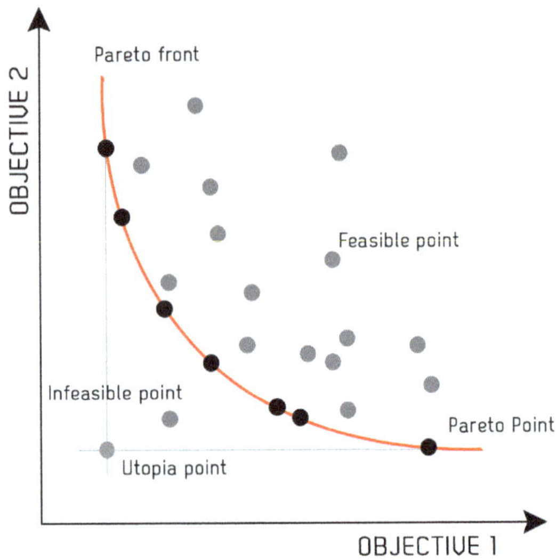

Figure 4.29 Pareto-Front (red) for a two objectives optimization problem.

Multi-objective optimization is applied in many areas, such as engineering, economics, and logistics. The optimal decision must be made in the presence of compromises between two or more conflicting objectives. For example, this could include minimizing the costs of car manufacturing while simultaneously increasing comfort or ensuring absolute safety while minimizing material usage.

4.5 Topology optimization methods

Topology Optimization (TO) is a computer-based calculation method. It is used to determine a favorable basic form (topology) for components. Topology Optimization is based on using no excess materials in component construction. For a basic initial shape under a set of given loads and boundary conditions, the optimization process searches for an optimized material layout within a given design space that meets the performance objectives. These objectives can be structural behaviors or constraints. For example, a solution with the highest possible stiffness while maintaining a maximum weight (Figure 4.30) (Bendsøe & Sigmund, 2004).

Other optimization objectives may include minimizing stresses, increasing eigenfrequency, or reducing mass.

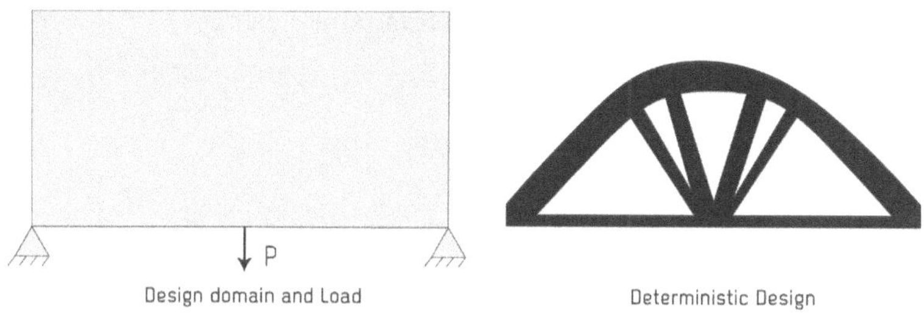

Figure 4.30 Form optimization of a beam under given loads and boundary conditions.

A general mathematical formulation for an optimization problem can be described as (Equation 4.2).

$$\text{Minimum: } f(x,y) \tag{4.2}$$

where:

x = parameter or variable
y = reaction function based on x

subject to:

the design space of x
the restriction functions.

SIMP method stands for Solid Isotropic Material with Penalization. A rigid structure is a system that shows the least displacement under certain boundary conditions. This displacement is caused by the strain energy that is subjected to the structure. The lower the strain energy, the higher the stiffness of the structure. Accordingly, the objective function here is to minimize the strain energy (Equation 4.3).

$$\int_{\Omega} \emptyset\,(p)\frac{1}{2}\,\sigma\varepsilon\, d\Omega \qquad (4.3)$$

where:

- p = the discrete selection field, which describes the material distribution in an optimized form. Theoretically, p refers to all points in the structure. These points either have value of 0 (removed) or 1 (kept).
- Ø(p) = an auxiliary function that describes the strain energy
- σ = stress
- ε = deformation
- Ω = the function that has to be minimized

The more massive the structure, the lower deformation that will occur. Whereas more material means higher cost. Therefore, the optimization has now a new volume restriction. Back to Equation 4.1:

$p \in \{0,1\}$ (0 or 1)

$\int_{\Omega} d\Omega \leq V$ (boundry conditions)

Having a discrete value of 0 to 1 for all points in the system makes the implementation of certain mathematical functions not suitable. Therefore, the assumption of linearity is taken to implement the problem in finite element analysis, where p varies linearly over a domain of (0,1) that allows it to be programmed:

$p \in \{0,1\} \rightarrow p \in [0,1]$ (linear)

The goal is to define p, which will help to satisfy the equation.

A simple SIMP example is when a simple cantilever system is assumed. The system (design space) represents the discrete selection field with black and white pixels. The Black pixels correspond to the dots in the body, where the discrete selection field p = 1 and the materials are kept. The white pixels are where the discrete selection field p = 9 and the material is removed. The grey zone represents all values that are $0 < p < 1$, a state between retention and removal (Figure 4.31). After running the calculations, the final form is generated based on the defined Rejection Ratio (RR). The Rejection Ration has to be accurately calibrated to grant the accuracy of the optimization (Tedeschi, 2014).

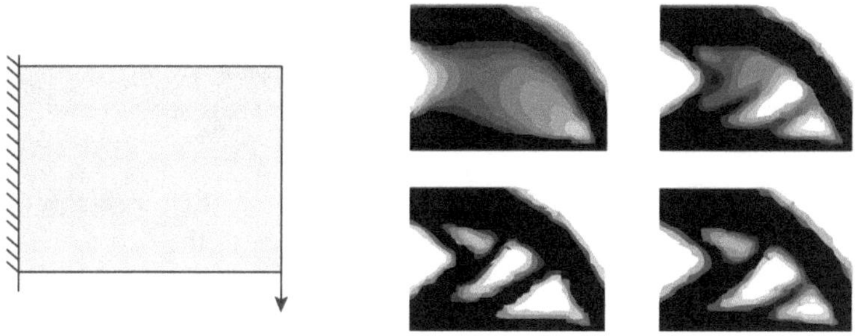

Figure 4.31 Iterations of SIMP-method.

Different other methods had been developed to solve TO problems such as Evolutionary Structural Optimization (ESO), Bi-directional Evolutionary Structural Optimization (BESO), Topological derivative, and Level-set-method (Rozvany, 2009).

Topology optimization has been evolved and becomes a practical design tool. This method was first applied to mechanical elements. These elements have proven success in automotive and aerospace applications. Accordingly, topology optimization was successfully introduced to other industries. The rapid development of Additive Manufacturing and the ability to manufacture complex topologies helped the integration of topology optimization into these industries.

4.6 Design for Manufacturability and Assembly (DFMA)

Design for manufacturability and assembly (DFMA) is the consolidation of product design and process planning. The engineer is often not aware of design impact on manufacturability. He designs parts only for functionality. After design release, the manufacturing engineer starts to work on the feasibility review and the manufacturability study. The sequence of the process does not permit trade-off analysis during the early stages, and it is time-consuming. 70% of the manufacturing costs of any product such as material cost, processing, and assembly are based on design decisions, whereas only 20% are based on production decisions such as process planning or machine tool selection. Accordingly, it is advisable to evaluate the impacts of the design specifications on manufacturing and assembly (Chang & Wysk, 1997; Kamrani & Nasr, 2010).

DFMA is an analytical process that results in a group of design principles and guidelines. This group of design principles and guidelines consider different aspects, such as design, development, manufacturability, assembly, total parts, and modularity to reduce costs and manufacturing complexity. DFMA has been applied in several industries to enhance products. These enhancements are in different areas such as cost, quality, manufacturing, and reliability.

DFMA focuses on simplifying the parts while evaluating assembly. The process works on all possible enhancements that could be done to the part to improve several aspects such as robustness, cost reductions, and the overall design. DFMA is a systematic evaluation process that is advisably applied during the early design phase. Design to manufacture analyzes the geometry of the part to define the suitable manufacturing process and its impact on the material and cost. Design to assembly analyzes existing parts for matching and suitability. These two processes complement the final goal, which is designing a product that is easy to assemble and economical (Figure 4.32)(Chang & Wysk, 1997; Kamrani & Nasr, 2010).

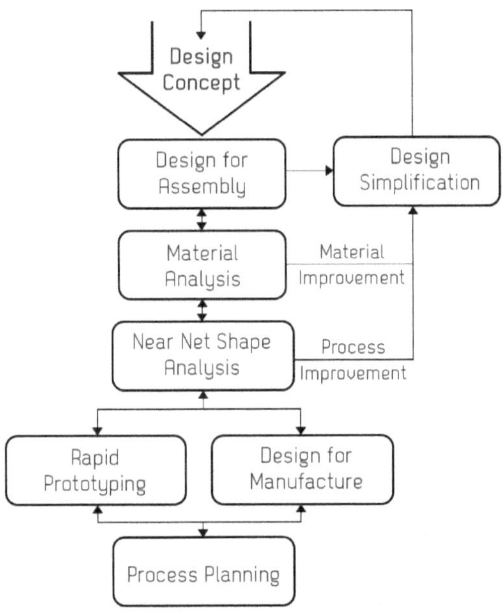

Figure 4.32 Scope of Design for manufacturing and assembly (DFMA) (Kamrani & Nasr, 2010).

There are general guidelines that help organize the DFMA process. Each is a stage where a particular aspect had been investigated to obtain possible improvements. These can be listed as a set of rules (Chang & Wysk, 1997):

1. Reduce the number of parts: helps reducing manufacturing costs and processing time.
2. Modular design: helps speed the manufacturing process and increases the possibility of cost optimizations.
3. Use of standard components: they are cheaper and more widely available on the market.
4. Design for multifunctionality: helps to reduce the number of parts refer to rule no.1.
5. Design for multiuse: parts are suitable for different systems.
6. Design for easy assembly: designing the optimum combination of manufactured elements to reduce any extra operations.
7. Avoid separate fasteners: fasteners increase manufacturing costs.
8. Minimize assembly instructions: assembly can be carried out from one side if possible.
9. Minimize transportation.
10. Increase certification levels.

These are general rules that may be followed during the process. It is not always possible to achieve all aspects. However, it is possible to apply as many as possible based on the required design/product.

In the case of designing façade node, it is essential to focus on a) reducing the number of elements of the part, b) easy assembly, and c) minimizing assembly instructions.

The reduction of the number of elements eases the manufacturing process and speeds up the certification processes. The easy assembly and the minimization of assembly instructions ensure that the product can be used with any kind of labor, which opens new markets.

5 Proof of Concept

The integration of complex free-form surfaces in architecture presented considerable challenges to the fabrication and construction of these structures. To be able to identify the planning concept, which is the purpose of this dissertation, the sequence of the process and the requirements of each phase had to be clarified.

To grant freedom of design, it was determined essential to design a system that is not geometry related. This is presented in the N-AM 1st generation, where a proposal for a parametric node was presented. The proposal presented a particular node structure that could be parametrically oriented and could be applied with standard façade systems. To obtain more freedom, the decision was made to move to the file-to-manufacture concept and to utilize parametric design for the generation of the nodes. This was tested in highrise-tessellation, where different tessellations were tested against a free-form surface. Voronoi-façade was the phase that improved the generation process of the nodes and that tested the first approaches to optimize the façade, looking for the lightest structure possible. At this stage, planar surfaces were not always guaranteed. Accordingly, in paneling-free-form, many tessellations were generated based on a free-form surface, and the optimization to reach maximum planarity was tested. For the grid Voronoi Shell structure, the structural performance and arrangement of the cells were optimized to reach best structural performance, the nodes were topology optimized based on the subjected load cases, and the influence of connection type on topology optimization was investigated.

Physical prototyping was essential to test the concept and the knowledge collected through the digital prototyping/experimenting phases. Holger Strauß in his Dissertation "AM Envelope - The potential of Additive Manufacturing for façade construction" presented Nematox – one of the first additively manufactured aluminum nodes for standard façade system. In N-AM 2nd Generation was a test of the first version of a generation/optimization script that allows generating the connecting node in the topology optimization form based on the structural analysis in one running programming script. N-AM_Li3 is the final version that shows an example of applying all aspects that were investigated during this dissertation. This example is designed and engineered to fulfill all requirements that are needed to implement the planning concept (parametric design and additive manufacturing) in the façade industry.

The concept has been integrated into teaching module "Façade Technology 1" at the Technical University Darmstadt, and has been applied in many Bachelor and Master research projects. The idea was to investigate the potential of the method and test its acceptance by architects and engineers.

A. Mohsen, *Design to Manufacture of Complex Building Envelopes*,
Mechanik, Werkstoffe und Konstruktion im Bauwesen 56,
https://doi.org/10.1007/978-3-658-30204-7_5

Different software was used during the research such as Rhinoceros, Grasshopper, solidThinking Inspire, Solidworks, ANSYS, and Rstab. Each was responsible for a specific process. These processes combined helped the planning concept formation.

5.1 Digital Prototypes

5.1.1 N-AM 1st Generation

At this early stage of the research, the implementation of parametric modeling/design was the target of the investigation. Additive manufacturing was not considered. A spherical shape is a precise round geometry in the three-dimensional space; any object running in parallel to its surface could move a full angle in the X, Y, & Z axes. Thus, having a ball connection integrated into a node will be sufficient to provide enough freedom for the connecting profiles. Different designs were proposed as N-AM series (Figure 5.1).

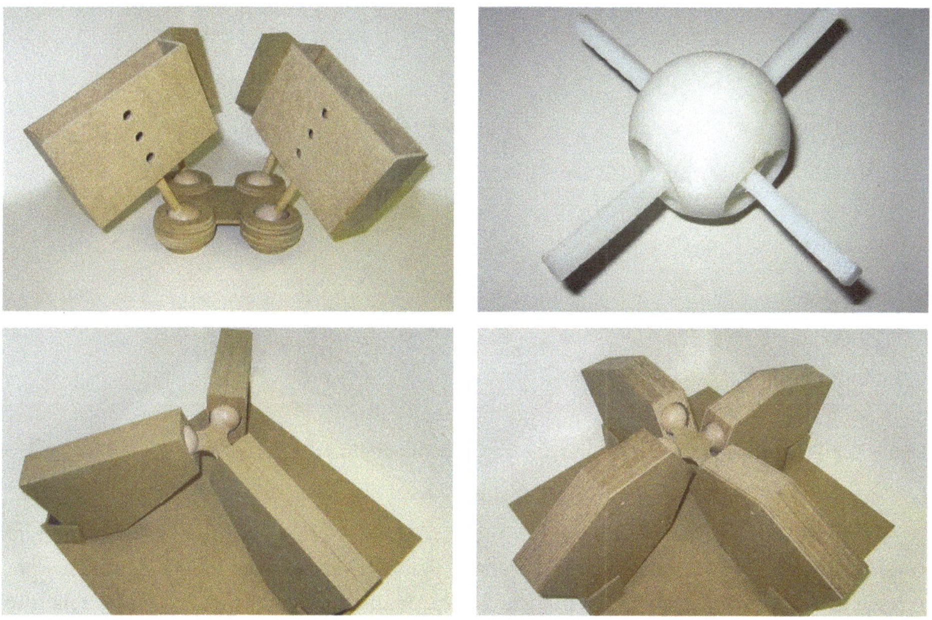

Figure 5.1 N-AM series prototypes.

The node design was defined, including: Node core, which is connected to ball shape that its count is defined by the number of connecting members/profiles; Ball connection, which is connected to the profile insertion; and profile insertion, which will be fixed to the profiles at the end of the installation.

After defining the design of the façade unit, the orientation of the profile has to be defined. The orientation of the profiles will generate the rotation angles that the node has to accommodate. To generate the process parametrically, the node and its connecting elements were digitally/parametrically modeled. The model was then integrated into an optimization algorithm. The objectives of this algorithm are as follows: (1) to find the normal vector of the profiles and orient these profiles accordingly, (2) to align the node with the profiles, (3) to find the optimum center of the node, and (4) to locate the node as near as possible to the top corner of the axes (Figure 5.2).

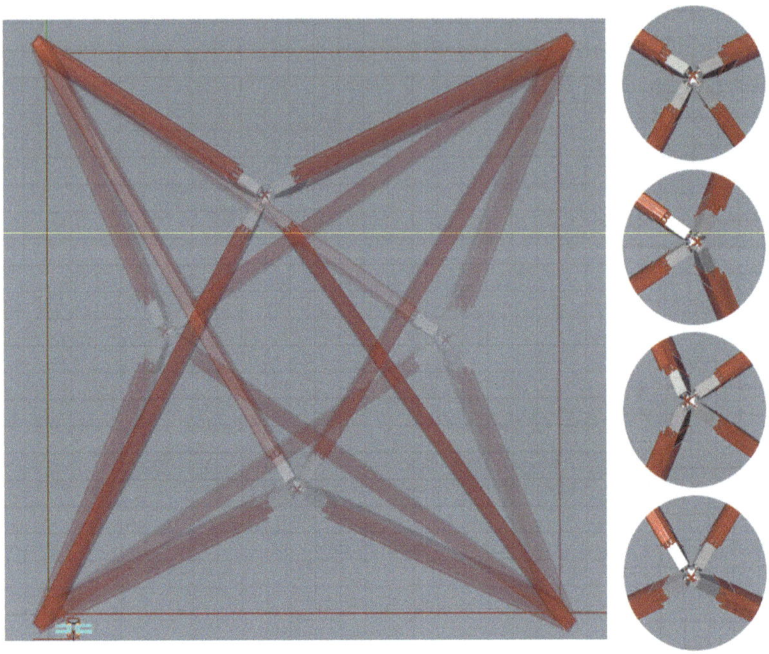

Figure 5.2 Different locations that the node can accommodate.

The final version of N-AM aimed to increase the structural stability of the node and to refine its design (Figure 5.3). The 1st generation of N-AM proved its goals, which were to have the flexibility to form sophisticated façade units parametrically and to generate the geometrical information that is needed to manufacture and assemble the façade unit.

The problem was at the basic geometry of the façade element. The node could be only accommodated when the basic borders of the façade element is a regular shape. Once the ratios of the basic borders move toward irregularity, the node can't be correctly positioned. In addition to that generating a clean fit cut between the profiles become impossible. The decision was made to digitalize the node as a whole and to start integrating additive manufacturing into the process to overcome these geometrical constraints. Highrise-tessellation was the next step, in order to investigate the formation of grid shell structures based on free-form surfaces, and to analyze the generation of the nodes.

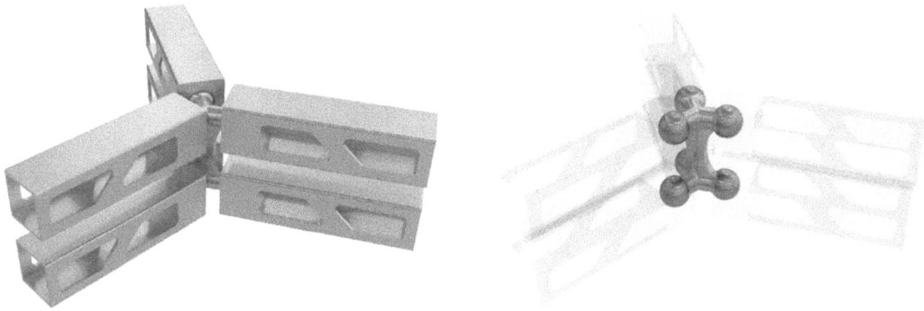

Figure 5.3 N-AM 9 renders.

5.1.2 Highrise-tessellation

The idea was to test different approaches to the subdivision process for given free-form surfaces. Simple tower geometry was generated, and the tessellation process was carried out. For the tessellation process, some geometrical requirements had to be defined. These requirements were as follows: the maximum length of the profiles was limited to 3 m, the minimum length of the profiles was limited to 0.9 m, and the minimum achievable angle between the profiles was limited to 30°.The tessellation was considered for triangular and quad meshes only.

For the tessellation types, the following were used: (1) a diamond shape that is subdivided into three polygons (quadrilateral), (2) a compound form of triangular and squares, and (3) a pure triangular mesh that forms a pyramid shape, (Figure 5.4). The main objective was to transform these tessellations to cover a free-form surface with straight bars/profiles and planar surfaces.

The three options both contained planar surfaces and straight bars/profiles. The first variant was generated by forming polygons. Based on their structural nature, these polygons need a bracing system to stabilize the overall structure. The second variant was generated by mixing triangular and square units in a regular pattern over the free-form surface. The triangular elements could act as bracing units to the whole structure. The third variant presented the possibility of dividing the uneven diamond shape into triangular surfaces (Figure 5.4).

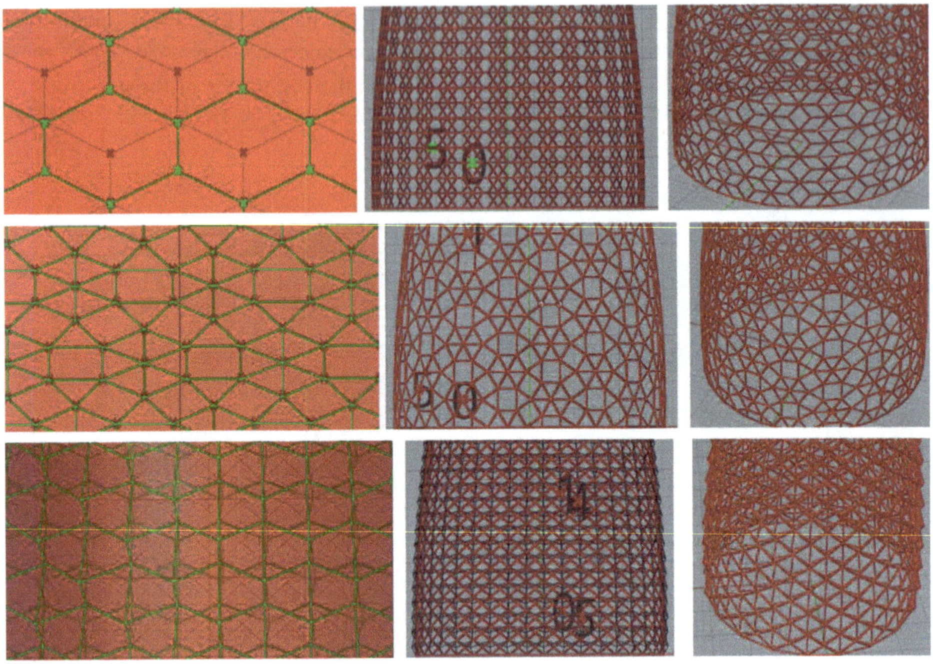

Figure 5.4 Different tessellation types.

All three variants contain different bar lengths and irregular angles. Each variant had two different types of nodes, which were congruent. The node in variant 2 had to be connected to five bars/profiles, and the node in variant 3 had to connect to eight bars/profiles. These nodes were challenging because of the huge number of connected elements, and the complex load condition that would occur (Figure 5.5).

Round profiles were considered, and a simple parametric model for the nodes was generated. It was clear that based on the free-form surface curvature, the profiles tended to have different lengths to accommodate this curvature.

The process of the grid shell structure generation was clearly defined for standard symmetrical shapes. Modern architecture tends to complex forms and irregular shapes. In the context of the Voronoi-façade, the investigation was directed to explore free-form complex shapes and irregular grid shells and to analyze the node generation for rectangular profiles.

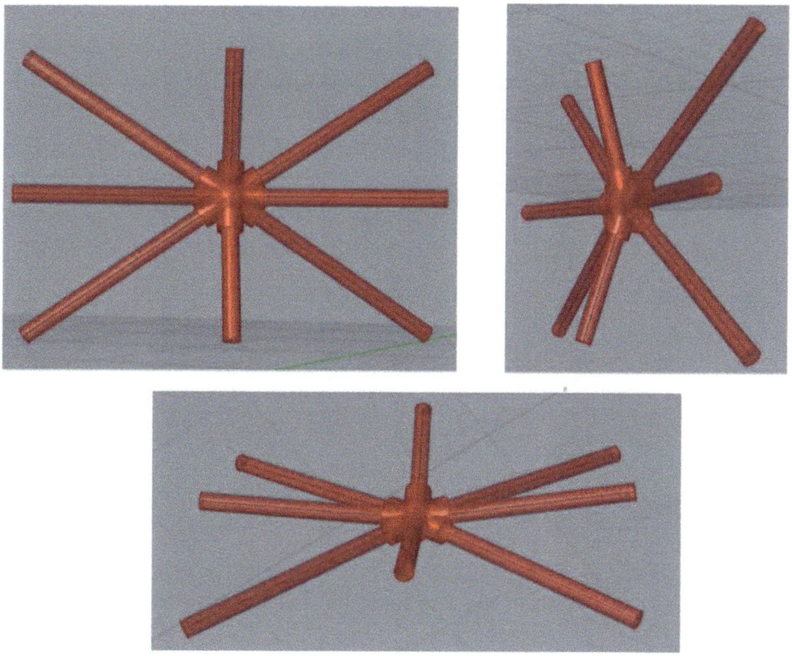

Figure 5.5 8-arm Node point.

5.1.3 Voronoi-façade

Voronoi diagrams are a way of dividing surfaces. It is named for the mathematician Georgi Feodosijewitsch Woronoi. These diagrams encode proximity information to help find which objects are closest to which certain points (The nearest neighbor rule) (Okabe, Boots, Sugihara, & Chiu, 2009).

Voronoi-façade was developed to investigate the subdivision of free-form surfaces. A free-form surface was defined, and the subdivision of the surface with Voronoi tessellation was generated. The Voronoi cells were defined through a random points population that was spread over the free-form surface. The location of these points on the surface could be optimized (changing location over the surface). The optimization objective was to find the lightest possible structure for this façade.

For the goal of the dissertation, it was essential at this stage to define the generation of the set of data that is needed for manufacturing. The line model of the façade was defined and based on load cases such as dead load and wind load, then the structural calculations were made. According to the forces that acted on the model, the cross-sections of the profiles were defined. The weight of the whole structure could then be calculated, and the optimization to find the lightest weight related to the distribution of the points over the surface could be run. After several iterations, a form was generated. This form represents the best distribution of points based on the system boundary conditions. The form is translated as a digital representation of the beams and the nodes with their related information -number, length, and form- (Figure 5.6).

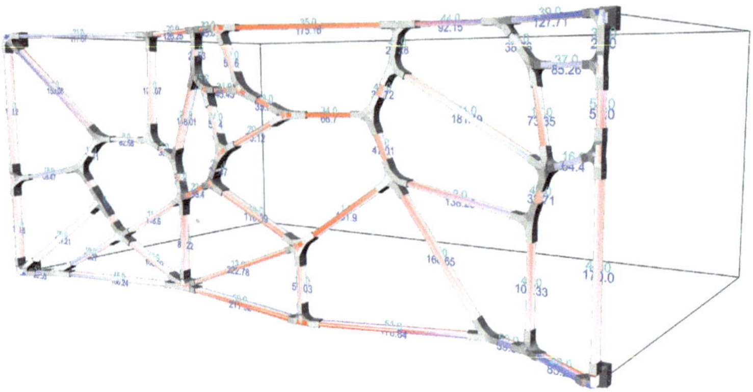

Figure 5.6 The digital model contains the data set of the structural elements of the façade.

Individual nodes were generated, which made traditional manufacturing techniques unsuitable. The integration of additive manufacturing to the research was the step to provide the nodes with more freedom, and to widen the range of design for the whole concept.

The model was scaled down and built with wooden sticks and 3D printed plastic nodes. The 3D printed nodes needed to fulfill certain aspects to be suitable for 3D printing. Closed polysurfaces are essential for a successful 3D print. A relation between the profiles had to be generated to form the boundary conditions that define the geometry generation of the nodes. The nodes were generated as a solid volume that contained the connection to the profiles. This connection was designed as a stick system, where the profiles were accommodated in a groove at the nodes. The process proved the validity of the set of data to generate assembly information. All related information for the nodes was present, and the 3D printing/prototyping of the models was successful (Figure 5.7).

The challenge was the generation of double curved surfaces between the profiles. The planarity of the surfaces was not one of the objectives of the optimization, and the random distribution of the points did not consider its geometrical relation with its congruent. Accordingly, Paneling-Free-Form was the concept to investigate the planarity of the surfaces as an optimization objective.

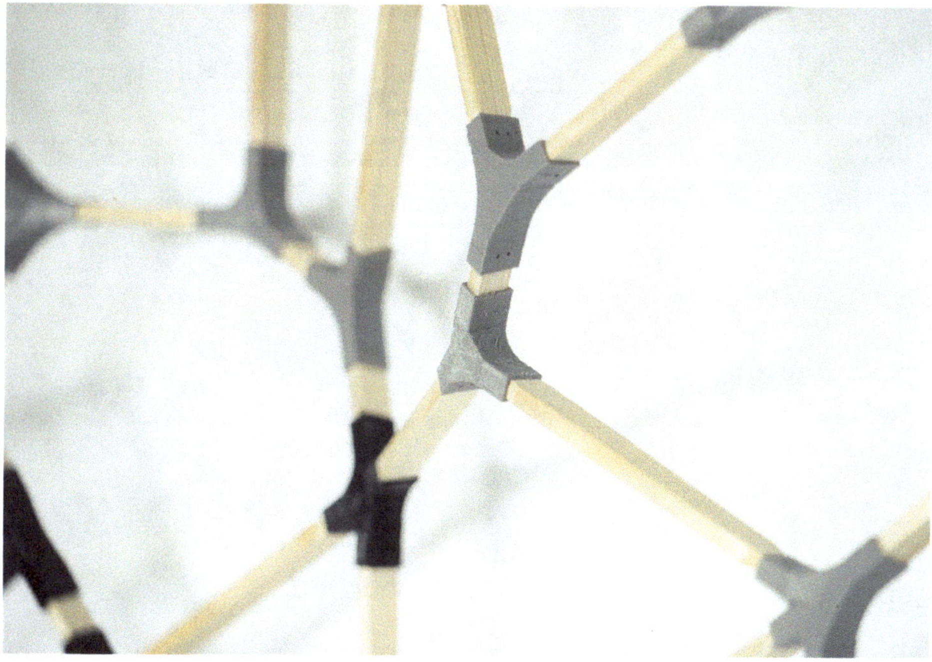

Figure 5.7 Prototype of the Voronoi-façade with 3D printed nodes in PLA material.

5.1.4 Paneling-Free-Form

Different types of curvature need to be examined to identify which façade shapes could be optimized. For this project, four different surfaces were designated to be the optimization subject for the process. These surfaces are one single curved surface and three double-curved surfaces. The single curved surface is a partial area of a cylinder. The three double-curved surfaces were a dome surface, saddle surface, and a merge between dome and saddle surfaces. These surfaces represent a façade for a two-story height building.

Three different types of tessellations were used. These types were two platonic tessellations (4-4-4-4, & 6-6-6), and Archimedean tessellation (3-4-6-4). To model the tessellations, a plugin called "LunchBox" was used. This plug-in can be integrated into the software Grasshopper. This plugin offers to project certain tessellations on a reference surface. A projected tessellation consists of curved edges and curved surfaces; their curvature is defined through the principal curvature of the reference surface.

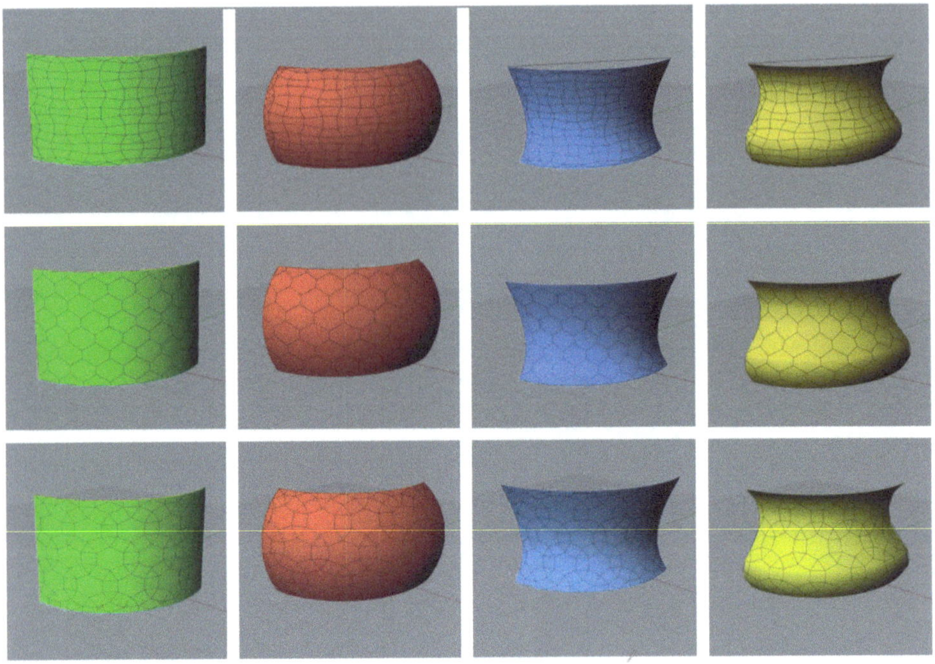

Figure 5.8 Single-curved surface (green), Double-curved surfaces (dome-red, saddle-blue, & dome saddle mix-yellow).

The complexity of the tessellation defines the computational performance needed to solve the optimization problem. The optimization process is highly nonlinear because it includes the planarity of quadrilaterals as an optimization objective. Helmuth Pottmann in "Geometry of Architectural Freeform Structures" said: "Experiments confirm that it is hopeless to optimize arbitrary meshes towards planarity, ..., However, optimization typically succeeds if initialized from a conjugate parametrization $f(u,v)$". In theory, the generation of an approximate segmentation to a reference surface is possible. However, there are different other constraints that influence the optimization such as minimum angles between edge surfaces (Pottmann, Schiftner, & Wallner, 2008).

The goal of the geometry optimization process is to transform all façade panels from the curved state into the planar state. The primary boundary condition is that neither the overall shape nor the tessellation for this shape should significantly differ from the reference surface. The challenge is to define the boundary conditions and the optimization parameters in a precise way to allow the optimization process to be run. These conditions and parameters could be defined in a way that allows the optimization to generate an approximate planar segmentation of the reference surface. However, the final selection of the optimal solution is the responsibility of the designer.

The angles between the individual panels cannot be optimized, as any change in these angles will result in a major change in the basic shape of the façade. The angles will be checked locally after the optimization process to check the profiles' orientation.

Spring-mass system, which is based on the assumption that each point has position, mass, velocity, and no volume in space, is applied to each node point on the reference surface. These node points are then computed as an input value into the so-called Kangaroo solver (Live physics engine for interactive simulation, form-finding, optimization, and constraint solving). As a result, a force vector is calculated for each node point. In order to calculate the new position of the points, Newton's second law, which states that a force (F) can be calculated as a product of mass (m) and acceleration (a), is applied (Equation 5.1).

$$F = m * a \tag{5.1}$$

The acceleration now can be calculated, as both force vector and mass for each point are known. Using the resulting differential equation of the traveled distance over time, the new position of the point can be calculated. These steps are repeated unless the designer terminates the optimization or a certain number of iterations is defined and reached.

As a second step of the optimization process, "Galapagos" component (Evolutionary solver) is used to achieve another optimization objective. This objective is to maximize the number of planar panels. The results are a set of solutions that meets the planarity goal, and they slightly differ from each other. The designer makes the selection of the geometry.

To break down the modeling of the optimization process, the steps that were followed have to be explained. All the node points of the model are defined as anchor points. This allows the reference points to resist movement. This movement is related to the weight of the points. The weight of the points is parametrized between 0 to 1 to allow further processing.

In order to preserve the basic shape and to avoid any significant changes in form: (1) the points lie at the bottom, and the top of the façades provided a high weight, (2) the points are only allowed to move along the top and the bottom of the geometry boundary lines, (3) and all the nodes are tested to ensure that they are laying on the reference surface. However, the weight of the points is kept variable to allow slight tolerances and widen the solutions range.

The length of the façade profiles was limited between 1,5 and 3 m to avoid customized profiles. To retain the basic tessellation appearance, the length of the profiles was limited to 10% longer or shorter than the reference profiles' length. The angles between the panels should only differentiate between 10, 5, or 1% in some cases. It is possible that the angles between the panels have to be entirely restricted (0%) to reach certain optimization goals. The resulted surfaces are tested for planarity and represented as the end form.

At this stage, all the boundary conditions and the first optimization stage are ready. The evolutionary solver "Galapagos" is now used to vary between the parameters trying both to increase the number of planar panels and to generate the final set of solutions (Figure 5.9).

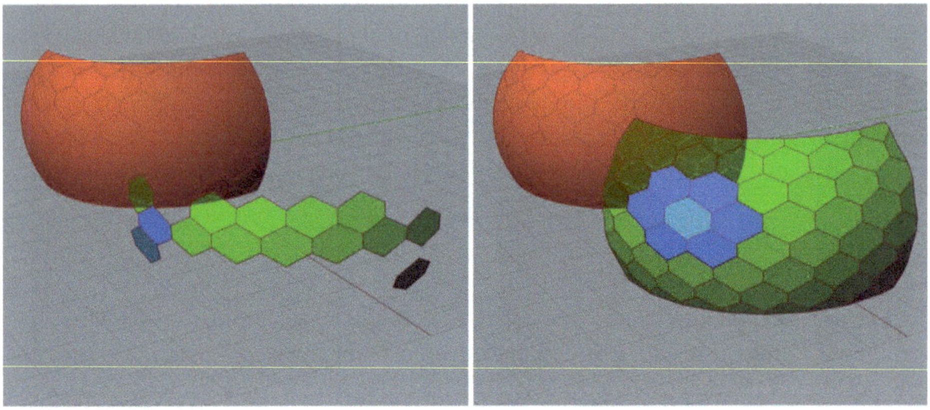

Figure 5.9 Optimization process through the Evolutionary solver "Galapagos."

For the first tessellation (4-4-4-4), the optimization yielded acceptable results for all the four façade surfaces. The set of solutions offered many valid combinations of parameters that ensured the planarity of the surfaces. For the honeycomb tessellation (6-6-6), only the dome surface yielded acceptable results. The other three surfaces tended to have a dramatic change of form. For the last tessellation (3-4-6-4), the triangulated elements helped the planarization process, however for the saddle surface and the mixed curvature surface, the optimization threw an error. This error was mainly because of the Gaussian curvature of the surfaces, which led to increasing the distance between the points that made the optimization not possible (Figure 5.10).

The generated surface can be selected at this stage, and the process to model its structural elements is done. The generated set of data was organized in a way to help to provide all related information for manufacturing and assembly.

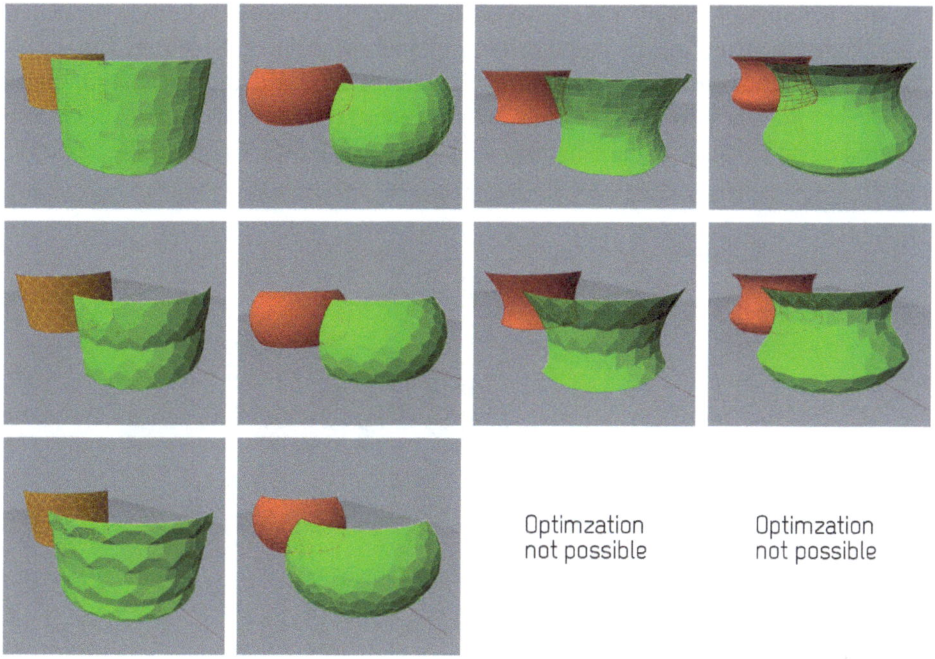

Figure 5.10 One case of the set of optimization solutions for each different tessellation/surface options.

To be able to evaluate the value of the optimization, a classification of the reference surface and the optimized surface had to be done. This classification defines certain aspects that are related to the model before and after optimization.

These aspects are:

1. Number of double-curved surfaces
2. Number of single curved surfaces
3. Number of the planar surfaces
4. Number of types for the double-curved surfaces
5. Number of types for the single-curved surfaces
6. Panel costs
7. Running meter of the profiles
8. Running meter of the straight profiles
9. Profile costs

The cost estimation is defined by the price on the market at the moment of planning, which must be known in order to calculate the reduction of costs based on the optimization process, as it is known that curved glass surfaces and profiles are costly.

After defining all related information for the structure, the nodes have to be generated and prepared for manufacture. Through the programming script of the project, it was possible to add a new stage that generates the orientation of the profiles, the inclination angles between the profiles and the glass panels, and the geometry of the nodes. The nodes can be additively manufactured in the same way as in the Voronoi-façade project (Figure 5.11).

Figure 5.11 Final geometry of one of the optimization results.

At this stage, the main principles of global structure and geometrical optimization process with one objective was clarified. This clarification is presented as Voronoi-façade with one objective structural optimization process to achieve a lightweight structure and at Paneling-Freeform with one objective geometrical optimization to achieve planarity of surfaces. The nodes for rectangular profiles were refined and stabilized. The next project aimed to investigate the optimization with two objectives (structural and geometrical) within the same structure and the transition from global optimization level to the local optimization level. The local optimization is presented as the topology optimization of the generated nodes and the improvement of the generation process of the nodes to match the requirements/rules of topology optimization, additive manufacturing, and the effect of different assembly types (bolted and welded). Global structural optimization of Grid Voronoi shell structure and topology optimization of its nodes investigated and clarified all these processes and their interactions to implement them into the planning method.

5.1.5 Global structural optimization of Grid Voronoi shell structure and topology optimization of its nodes

The goal of this project was to investigate the feasibility of evolutionary optimization on irregular building structures. A free-form three-dimensional structure was designed as coverage of narrow streets between three buildings. This free-form shape was the reference surface for the whole process. Genetic Algorithm (GA), which is based on an evolutionary optimization strategy, was chosen to link between the input (reference surface) and the output (optimized surface). Genetic Algorithms (GA) is a combinatorial heuristic technique that is effective and robust to solve multicriteria-discrete optimization problems (Wu & Chow, 1994). To perform the optimization process, Rhinoceros and Grasshopper software and its optimization engines -Galapagos and Octopus- were used.

The blueprint of the whole buildings is 30m x 40m, and it contains three buildings. The three buildings are separated from each other with narrow street respectively. The building height varies between 5m and 8m. The roof represents a free-form surface, which covers the three streets, and connects the buildings (Figure 5.12).

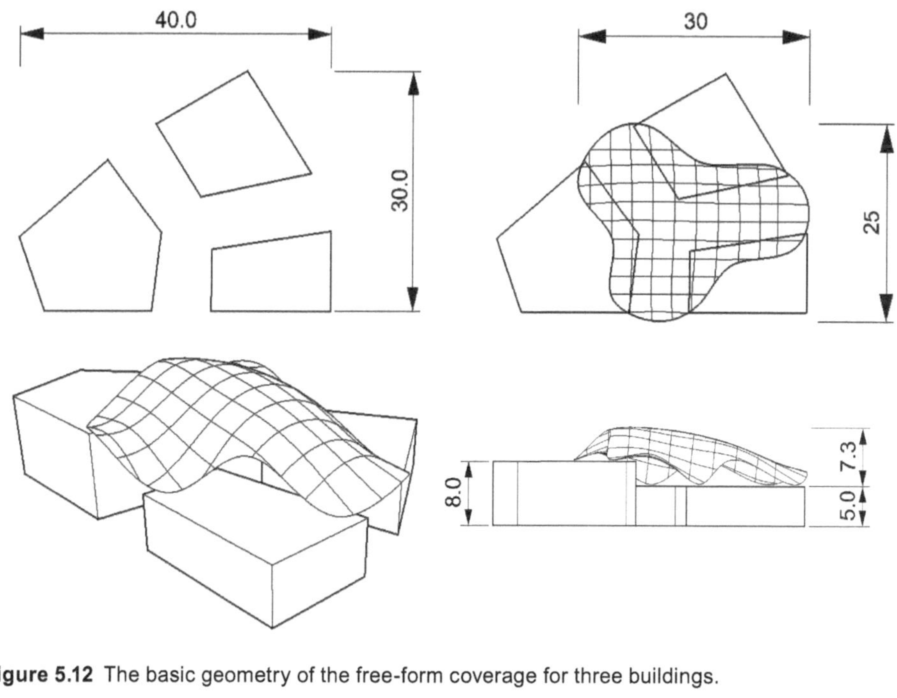

Figure 5.12 The basic geometry of the free-form coverage for three buildings.

Voronoi diagrams were used again for the generation of the single layer shell structure for this free-form surface. In most cases, they have exactly three bars/profiles that connect at a node, which will simplify the generation of the nodes. Voronoi diagrams can be controlled by a set of points, which results in more freedom for structure optimization.

Forty random points were populated over the surface. These points were the seeds to generate the Voronoi diagram that resulted in forty cells. However, the initial diagram was not yet suitable for structural analysis. The randomness of the points generated irregular cells. The length of the elements and the angles between them varied greatly. To bring the mesh to a regular distribution state, Lloyd's Algorithm was used. Lloyd's Algorithm is also called Voronoi iteration. It is an algorithm that generates an evenly distributed set of points. These points represent the subsets of Euclidean space that is transformed into uniformly sized cells (Lloyd, 1982). The algorithm ran in three steps; the points of the Voronoi diagrams were defined, the centroid of the cells was computed, and the points were moved to its cell centroid. After several iterations, the pattern was be regular (Figure 5.13).

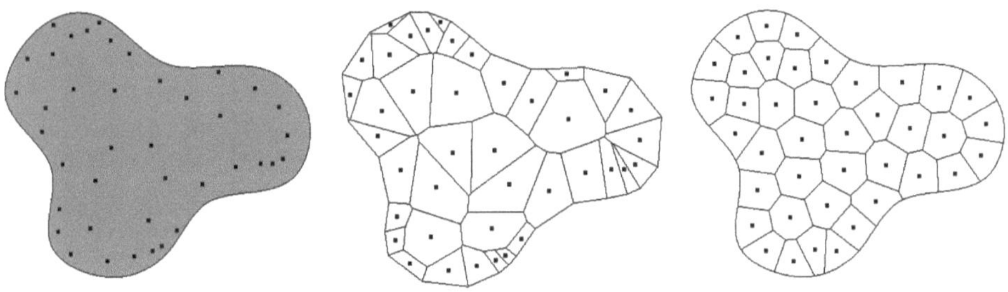

Figure 5.13 Lloyd's Algorithm relaxation of a Voronoi mesh.

This surface relaxation led to regular cell size, which had a positive effect on the manufacturability of the structure. This was done by reducing the length deviation of the profiles – from 7.3 m to 4.1 m, and the deviation of the surface of the panels – from 1 m^2 - 31 m^2 to 7 m^2 - 14 m^2.

The process was done on the 2D diagram. The regular pattern was then projected onto the Free-form surface to define its tessellation. The tessellation mesh represented the profiles that were oriented to the normal vectors of the free-form surface. The structure system was then defined and ready to be calculated. To start the calculation process, the support points needed to be defined. Two points were defined on each building as support points for the shell structure (Figure 5.14).

The following assumption was made: the structure consists of rectangular steel hollow profiles that vary between 40x20x3 mm to 400x200x12.5 mm, all nodes are rigid, and Steel S235 is material of the structure.

Self-weight, live loads, snow loads, and wind loads were calculated according to DIN EN 1991-1 (The Principles of Impacts on a Structure). The load combinations were defined and applied to the structure. As per DIN EN 1990 (Basis of Structural Design), the maximum deflection of the structure should not exceed 0.05 m. The static calculation for the whole structure was run.

An optimization cross section component (namely, an integrated optimization component in Karamba – a plugin for structural analysis for Grasshopper) defined the cross-sections of the profiles from the defined range of sizes. The cross-section optimization process was defined to generate a solution where the maximum utilization of elements is < 1.0, and maximum displacement is 0.05 m.

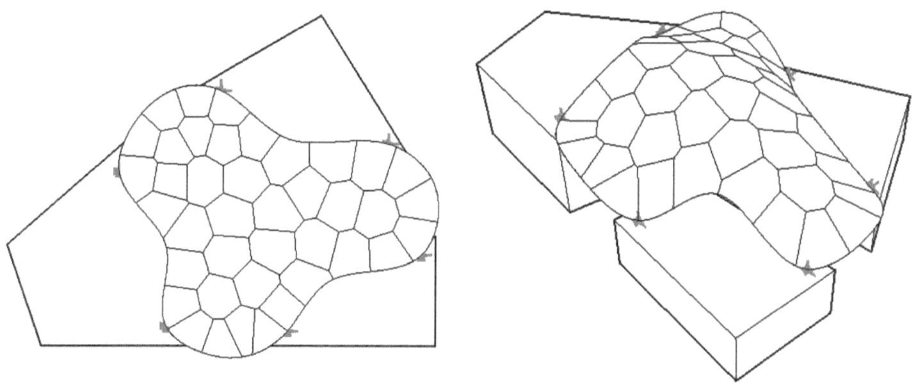

Figure 5.14 The tessellation of the free-form surface with regular Voronoi diagram, and the definition of the support points.

Now the parametric model of the structure has been built. All results after the static analysis were represented through the interface of Karamba. As the whole process was parametrized/programmed, there was a direct connection between the input parameters and the output parameters. The effect of any change in the input parameters on the end results could then be directly evaluated and visualized.

Two plugins, Galapagos and Octopus, were utilized to demonstrate the feasibility of the structural optimization strategy. Galapagos is an optimum finder for individual object functions that have only one target value (single-objective optimization). Octopus is a solver for multi-criteria optimization problems (multi-objective optimization). The operation sequence is a chain of processes that runs one after another (Figure 5.15).

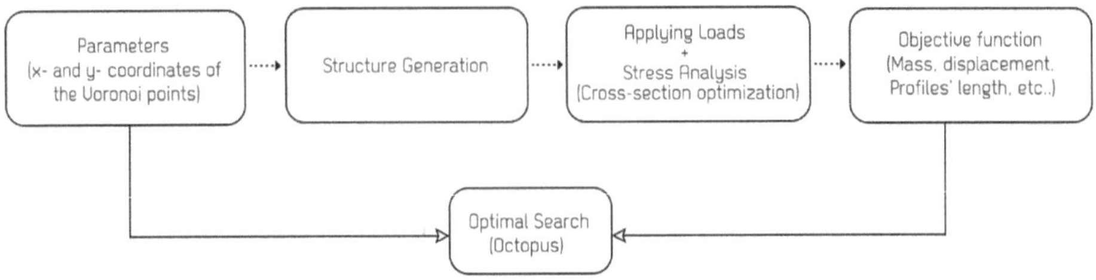

Figure 5.15 The sequence to formulate the structural model and its optimization process.

The optimization objectives were Mass: lightest structure possible, Displacement: minimum displacement possible, and Elements length: the length of the elements should lie between 1 m and 5.5 m.

To adjust the evolutionary algorithm, certain parameters had to be defined (Grefenstette, 1986; Haupt, 2000):

- Population size: number of solutions in a generation – a population that is too large slows the calculations. A population that is too small cannot cover the search space.
- Elitism: percentage of the new solutions that bred from the elite (preferred solutions) instead of the entire search space.
- Mutation probability: the probability that each will be mutated.
- Mutation rate: too high mutation rates reduce the performance of the evolutionary algorithm. Too low mutation rates mean small changes in the values of the parameters. It is recommended to have a mutation rate between 0.05 and 0.1.
- Crossover rate: the probability of two subsequent solutions to exchange their parameter values. High crossover rates are associated with low population size. A crossover rate between 0.8 and 0.95 is recommended.
- Number of evolutions: the program stops once this number is reached, and the most stable solutions are found.

The needed parameters for the optimization could then be defined (Table 5.1).

Table 5.1 Optimization parameters.

Population size	50
Elitism	0,7
Max Generations	200
Mutation Probability	0,1
Mutation Rate	0,01
Crossover Rate	0,8

The distribution of the solutions in the course of optimization shows how the solutions developed through the generations. At the first generation, the solutions were spread over the search space (yellow points). At the fiftieth generation, elite solutions that were bred from the last generation were defined (red points). At the hundredth generation, more solutions were distributed over the borderline. At the end of the two hundredth generation, a clear stable boundary that represented Pareto-front is generated (Figure 5.16).

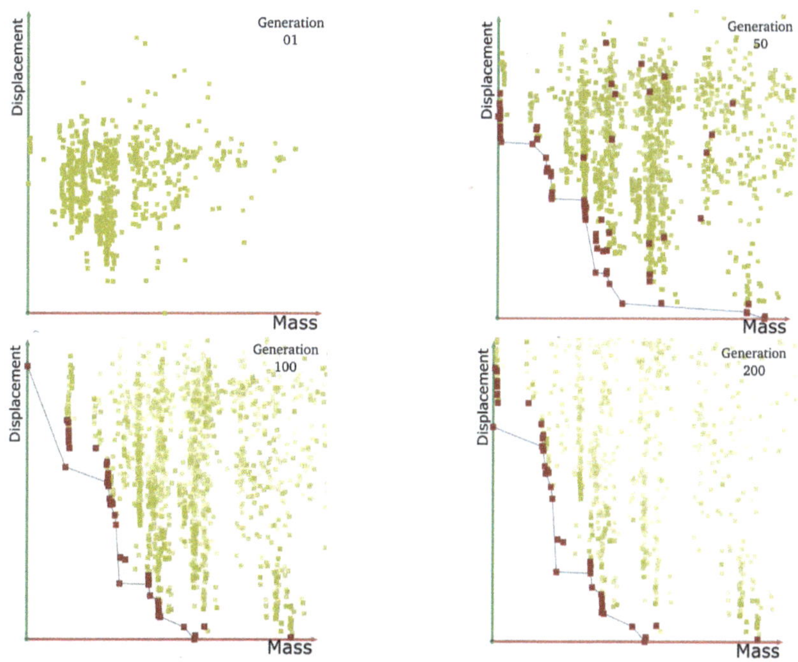

Figure 5.16 Solutions, respectively through generation 1, 50, 100, 200.

After two hundred generations, the Pareto-front was established. All points on the Pareto-front theoretically are optimized solutions. The final step was the selection of the solution. Four solutions (a, b, c, & d) were selected and compared (Figure 5.17).

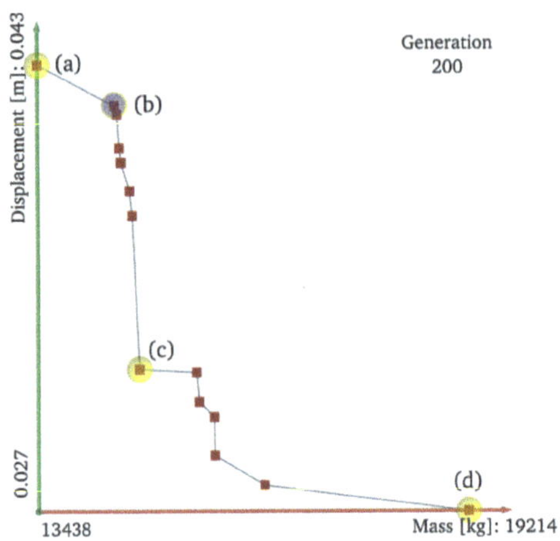

Figure 5.17 Pareto-front.

The related information of all four solutions was extracted from the data set generated with the optimized solutions (Table 5.2).

Table 5.2 Solutions properties.

	Mass [kg]	Displacement [m]	Min. Profile length [m]	Max. Profile length [m]	Cross-section (profiles network)	Cross-section (border lines)
a	13438	0,043	1,5	4,91	FRR 200x150x7	FRR 250x200x8
b	14491	0,041	2,7	4,92	FRR 200x150x8	FRR 250x200x8
c	14817	0,032	2,6	5,11	FRR 180x140x8	FRR 250x200x10
d	19214	0,027	1,2	5,25	FRR 200x160x10	FRR 250x200x12

For aesthetic and performance reasons, solution b was chosen. It had a balance between weight and deformation. Its profile minimum and maximum lengths laid within the manufacturing limit. Now the stress check for the solution could be done (Table 5.3).

Table 5.3 Stress design for the optimized solution.

	Min.	**Max.**
x-displacement [m]	-0,017	0,011
y-displacement [m]	-0,018	0,016
z-displacement [m]	-0,006	0,039
Total displacement [m]	0	0,041
Compressive stress [MPa]	0,31	224,62
Tensile stress [MPa]	7,92	247,76
Shear stress [MPa]	0,09	51,52
Utilization	0,035	0,996

The utilization of the elements was presented in ingredient colored scale. The color ingredient varied between 0% and 100% of the utilization of the elements (Figure 5.18).

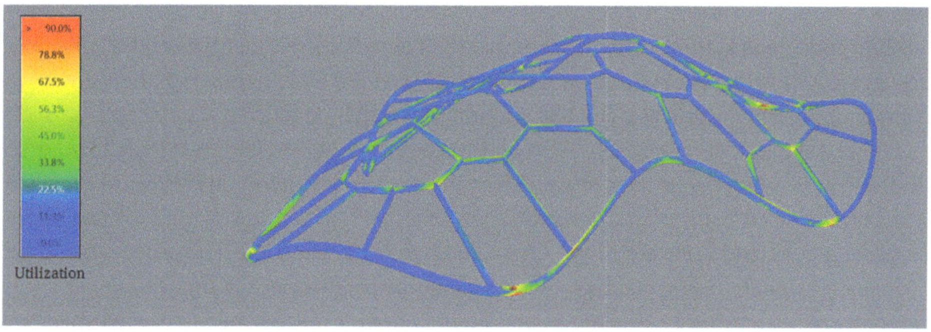

Figure 5.18 The utilization of the profiles for the optimized solution.

The whole structure was then defined, and the connection between the profiles was determined. Based on the size of the chosen/dimensioned profiles, the size of the node could be well-defined. To run the optimization on the local level, topology optimization (TO) was applied to the nodes. The generated forms based on topology optimization were characterized by complexity and irregularity. Additive manufacturing is able to manufacture such complex and irregular forms. The advantages that TO brings to the process are: the reduction of the weight of the node, which means the reduction of manufacturing costs, and the reduction of the weight of the whole structure, which means more durable/optimized structure design.

The node consisted of six different sides. Three straight sides were connected to the profiles. Then, three sides with rounded corners were connected to accommodate the geometry curvature, to grant the transition between the profiles, and to assure the flow of the forces (Figure 5.19).

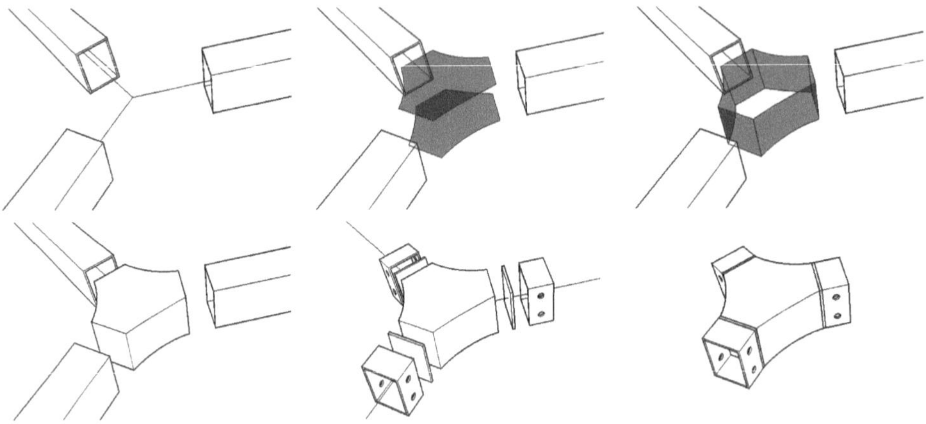

Figure 5.19 The Generation of the geometry of the nodes.

The node consisted of three main parts: (a) The node kern that was considered as a design space for the TO process, which was represented as a volume for the weight reduction optimization process. (b) The straight plates that connected the node kern to the node end, (c) and the connection between the node end and the bar end, which could be designed as a bolted connection or welded connection. The generation process of the node was parametrized using Rhinoceros and Grasshopper, which meant that all the nodes could be generated in one step. Yet, the calculation time was high. The process could be done node by node or in a batch of a moderate number of nodes at a time.

The basic shape of the nodes was generated. The loads on these nodes had to be determined. During the calculation and optimization process, the internal forces of all structural elements were generated along with the geometric information of the elements such as cross-sections and orientations. With this information, the loads on the nodes were determined. All nodes were numbered to give an overview of the whole structure, and to verify the exact node and all its information based on this numbering (Figure 5.20).

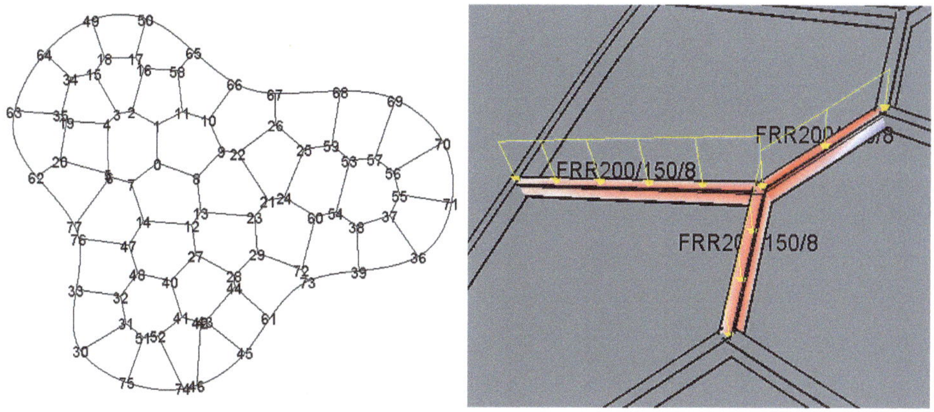

Figure 5.20 The numbering of all the nodes (left), Node 59 (right).

The number of the nodes was engraved on the joining edge surfaces of the node to simplify the assembly process. Node 59 was selected because of its high bending moments. The geometry of the node was generated, and the size was defined based on the measurements of the connected profiles (Figure 5.21).

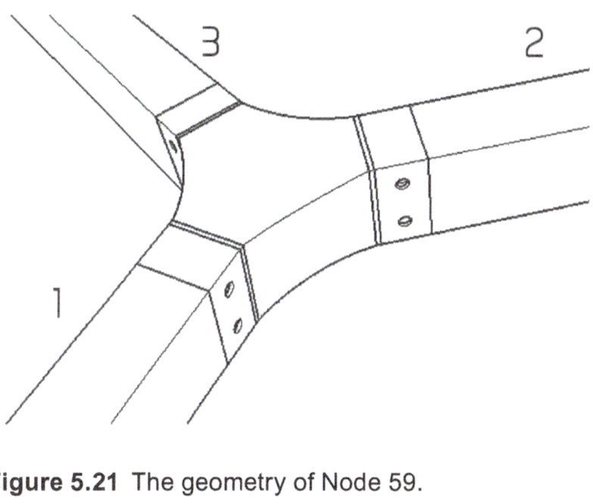

Figure 5.21 The geometry of Node 59.

All the internal forces were extracted out of the parametric model (Table 5.4). Then the topology optimization parameters were defined.

Table 5.4 Forces affecting Node 59.

	Position 1	Position 2	Position 3
Nx [kN]	-14,57	-25,83	-3,62
Nx [kN]	-14,57	-25,83	-3,62
Ny [kN]	-17,35	-2,31	1,52
Nz [kN]	-2,28	1,06	-3,31
Mx [kNm]	2,55	-3,72	0,00
My [kNm]	-15,42	-12,64	-6,69
Mz [kNm]	17,31	5,32	12,16

The node is considered as a solid steel component. A new topology was generated to design a light node with high stiffness. The node kern was defined as a design-space that was made of Steel S355, which is stronger than Steel 235 that was used for the profiles. It weighed 141.21 kg. The whole node with the not-design-space part weighed 160.49 kg.

The loads were then applied to the node. Two types of connections were considered, both welded and bolted connections. The stress distribution differed based on the connection type. Thus, the topology optimization was influenced. At the bolted connection, the bending moments were converted into force pairs (tension and compression), then applied to the bolt holes. To maintain equilibrium, one side of the nodes was defined as a support point. At the welded connection, the loads were spread over the cross-section and one side was defined as a support point (Figure 5.22).

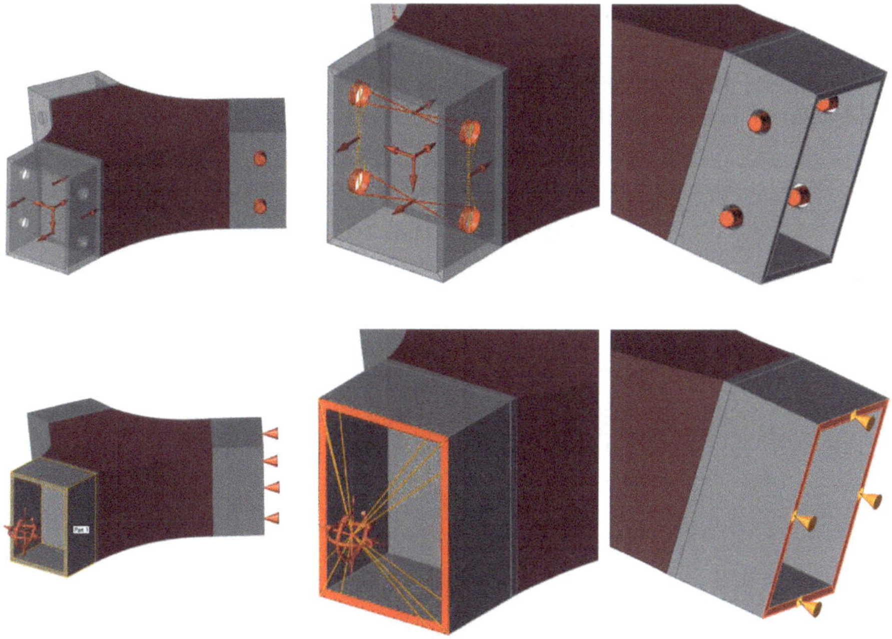

Figure 5.22 Boundary conditions of the node-bolted connection (top), welded connection (bottom).

There are two options for topology optimization (see 4.5). The first one is to maximize stiffness, and the second one is to minimize mass. For the maximize stiffness strategy, a mass target has to be defined. Accordingly, the search for the maximum stiffness is done under the given mass percentage. However, this cannot be achieved all the time. Some mass percentages will yield a result, but it is possible that these results exceed the yield point of the material. The process needs to be done iteratively to find the most appropriate mass percentage that can support the given loads. For the minimize mass strategy, the mass percentage is defined to represent the lowest weight that can support the applied loads. The permissible stress and deformation here were defined.

In general; stiffness maximization results in a shape that resists deformation, and mass minimization results in a light shape with greater deformation.

The settings for topology optimization were defined:

- Mesh size: 10 mm.
- Optimization Objective: Maximize Stiffness.
- Mass Targets: 30%, 20%, and 15% of the design space.
- Minimal thickness: 25 mm.

The topology optimization was calculated for the two different connection types (bolted and welded). At 30% and 20% mass targets, the results yielded logical shapes and values (Figure 5.23) (Table 5.5). At 15%, the generated shape was not connected at critical locations, which made the results invalid.

Table 5.5 Topology optimization results for the bolted and welded connections.

Node		Total Mass [kg]	Max. Displacement [mm]	Max. Mises Stress [MPa]
Welded Connection	100%	160,86	0,34	161,98
	30%	58,58	0,41	104,12
	20%	50,22	0,50	116,48
Bolted Connection	100%	160,49	0,36	285,40
	30%	61,12	0,42	350,53
	20%	46,79	0,51	369,28

It was clear that the connection type influenced the topology optimization. The project ended with examining the raw optimization results. It became advisable to remodel the results in the form of smooth self-supporting elements. The new model was affected by the chosen additive manufacturing technique. The new model used the topology optimization results as a base shape. However, it should be checked if it yields the same structure behavior as the raw optimization.

To investigate the remodeling process on the raw optimization, the table was the next experiment.

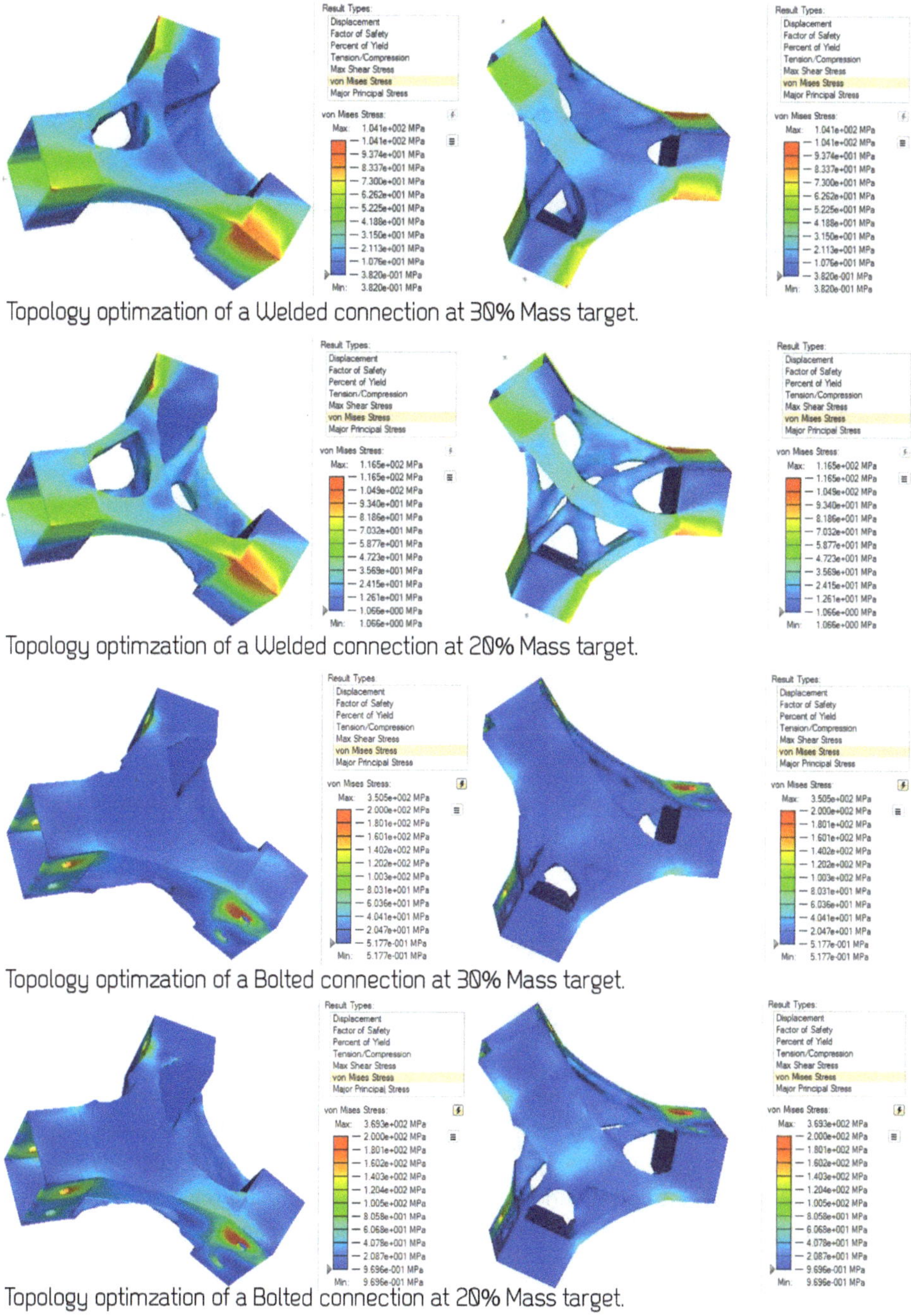

Figure 5.23 Topology optimization results at 30% & 20% Mass targets for welded (Top two) and bolted (bottom two) connections.

5.1.6 The Table

Figure 5.24 The Table – Topology optimized plastic nodes.

The table consists of a simple structure model that contains two rigid nodes in the three-dimensional space. This is defined by the table work area, four points that spread over the bottom of the table work area and allowed to slide all over it but not cross its side, and four support points that lays on the ground and have the same movement limitation as the upper four points.

200 kg weight was applied to the top four points, and the optimization searched for the best distribution of the points to reduce the displacement at the nodes. The form (Figure 5.24) was generated.

The reaction forces at the nodes were extracted to run the topology optimization. The topology optimization was done for polylactide material (PLA). The raw optimization was then remodeled to provide a smooth surface for 3D printing. It was obvious that the new model has to differ slightly with the raw model to reduce overhanging surfaces that require a support structure.

Further investigation at this stage was to design a unitized façade element and utilize the previous stages to test the feasibility of the collected pieces of information and to improve the whole planning method. In Façade Technology 1, a façade element was designed to test different optimization aspects. These aspects are related to predefined manufacturing constraints such as profile length, surface area, and nodes coordinates.

5.1.7 Façade Technology 1

During the teaching of the Façade Technology 1 course for the summer semester, 2018 at the Technical University Darmstadt, several projects were developed to try testing the planning concept.

The goal was to design a unitized element for a complex building façade design. The detailing of the element was simplified to match the planning effort with the time available during the course. Thus, wooden rectangular profiles were used, the size of the element was scaled to half, and the glass panel was exchanged for thin Acrylic glass.

One of the projects aimed to achieve the best user orientation, to have maximum contact with the outside, to provide noise scattering effect through adaptive surfaces orientation, and to maintain economical manufacturing value. Accordingly, a form that tried to accommodate all the project objectives was generated. The façade element consisted of a rectangular frame that contained nine profiles, which formed the glass surfaces. The profiles were connected at three interior nodes and four corner nodes (Figure 5.25).

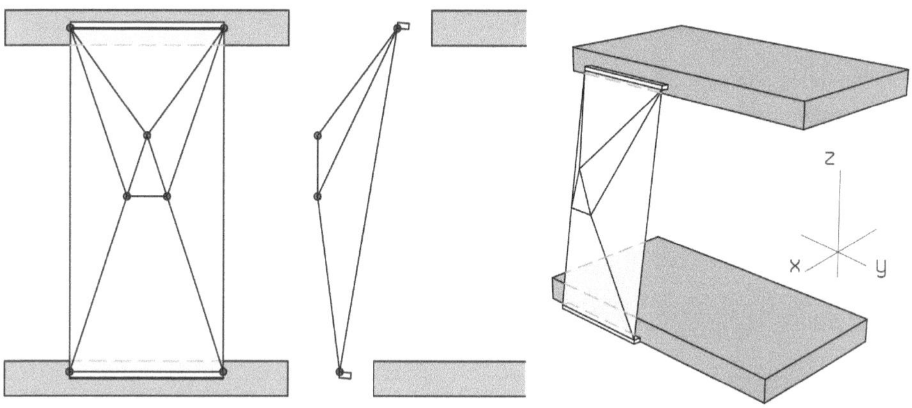

Figure 5.25 The basic shape of the façade element.

The optimization variables had to be defined. These variables were Z & X coordinates of the three free nodes, the length and coordinates of the profiles connected to the nodes, and the shape and area of the surfaces. Accordingly, three different optimization objectives were formed (same profile length, same surface area, and same profile length for certain group of profiles).

The first optimization objective aimed to yield nine profiles with the same length. The results achieved nine similar profiles with 730 mm deviation, and the form was not satisfying. The second optimization objective aimed to yield seven surfaces with the same area. The results achieved seven similar surfaces with 0.35 m^2 deviation. The last optimization objective was to define a group of the profiles that had to have the exact length. The results presented six profiles with the same length (deviation of 0,02 mm) (Figure 5.26).

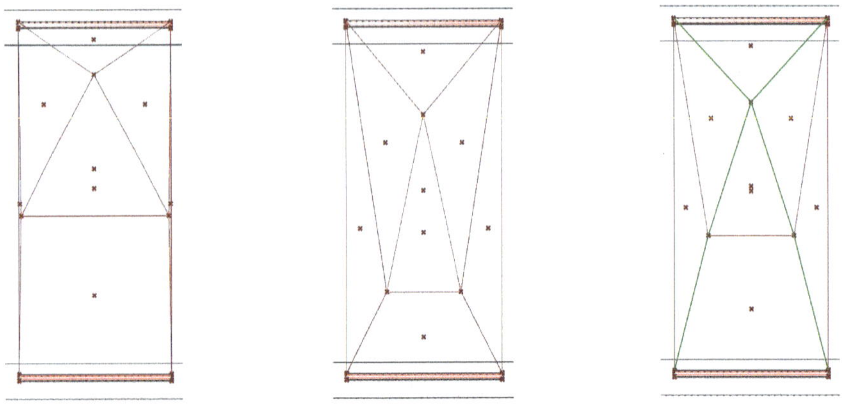

Figure 5.26 1st optimization objective results (left), 2nd optimization objective results (middle), and 3rd optimization objective (right).

The static analysis for the model was done according to DIN EN 1991-1-4, and the results were used to run the topology optimization for the nodes. The nodes were first designed. The node consisted of two parts connected to the profiles through a dovetail joint (Figure 5.27).

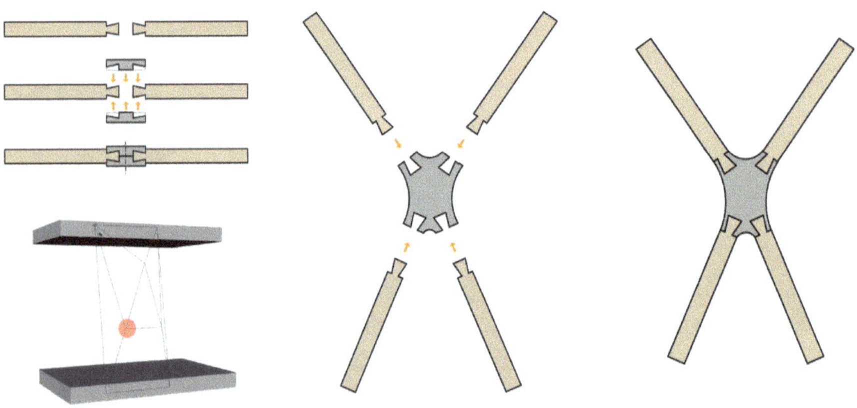

Figure 5.27 The node design using a dovetail joinery technique.

The two parts of the node were then fixed to each other using bolts. Now the design space for topology optimization was defined. Stiffness maximization was the concept applied for the topology optimization with 30% weight target, the minimum thickness of 4 mm, and steel as a manufacturing material. The node was remodeled using spline surfaces, which improved the 3D printing process. The raw optimization was used as reference geometry. The new model was simulated to compare its behavior to the behavior of the raw optimization (Figure 5.28).

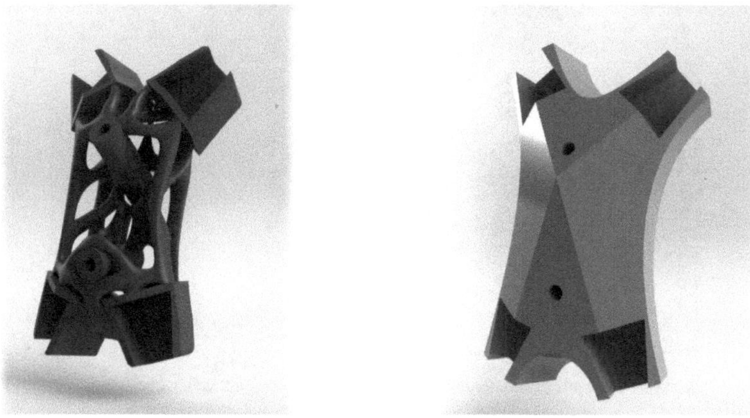

Figure 5.28 The form of the back-side of the node before and after topology optimization.

The façade unit was then built out of wooden profiles, acrylic glass panels, and plastic 3D printed nodes. The nodes were manufactured using an FDM (fused deposited modeling) printer (Figure 5.29).

All needed building elements are complete at this stage. The logical transition heads toward physical prototyping to move the concept from idea to a functional method and the node from prototype to a product. Therefore, physical prototyping was the next stage.

Figure 5.29 The designed façade mockup.

5.2 Physical Prototypes

Physical prototypes offer the transition of prototyping to actual product. It was crucial to test the acquired knowledge in producing a genuine part that could be tested and certified. The idea of integrating additive manufacturing into the façade industry is not entirely new. Several types of research have investigated this topic.

5.2.1 Nematox II

The research project of Holger Strauß at the University of Applied Sciences – Hochschule OWL, Germany, together with the Façade Research Group at the TU Delft and the industrial partner Kawneer-Alcoa, investigated the potential of AM for façade technology from 2008 to 2012. The result was the first free-formed aluminum façade node for an existing façade system (Stick-System).

In the field of AM, the technical development of the past ten years has been extraordinarily rapid and continues unabated. Developments in AM technology, new materials, quality assurance in the process, the simplification of software, as well as the increasing development of new ways of thinking for production with AM are ongoing aspects on the route to an established production method. Moreover, Holger Strauß's dissertation *"AM Envelope - The potential of Additive Manufacturing for façade construction"* shows that AM technologies enable new applications and developments.

The work aimed to fathom the potentials of the additive processes for a modified façade construction and to prove them on practical building components that meet building code regulations.

For this purpose, the technical possibilities of the AM methods were documented and evaluated in the context of façade technology. Based on various research projects carried out, relevant AM technologies for use in façades could be limited. The work shows that AM technologies offer great potential for the building envelope because architecture is a one-off with every new project.

From the evaluation of AM technologies, the requirements for the "printed" building envelope were formulated and brought into direct connection with the available AM technologies. The next steps in the development of an AM building envelope were described from a list of the required performance fields, but also via consideration of the historical development of façade technology. Strauß points out that every new development always has the potential for improved components as well as for improved AM systems technology and thus has effects in both directions of development – design as well as technology.

In order to connect the AM theory with the reality of the façade, AM prototypes were developed and produced in the research work, which has a proven relevance for their further development with AM. The prototypes made it possible to make the AM processes accessible to specialists from the building industry and comprehensible. The reactions from the industry consistently showed that resilient approaches have been developed and thus supported the core statement of the work that AM technologies have great potential for changing the construction methods of façades.

The façade node "Nematox II", in the version presented in this dissertation, is a real prototype. At first, it seems without any connection to a realistic production for a façade application. Nevertheless, it was further developed as a logical consequence of the first development steps with smaller components from a façade system, so thus offered a sound basis for its evolution with regard to the digital optimization and simulation of AM components.

In retrospect, the advantages developed can be seen from the prototypes which were not obvious at the beginning of the research work. The potential for improvement often only arises from the intensive examination of the standard used for a long time (in this case existing system components of the Stick-System): in the research work, ideas were formulated to show how the performative properties of the technology could be combined with an AM vision for the design. The results of this work were realized with a physical prototype – the "Nematox II" façade node (Figure 30).

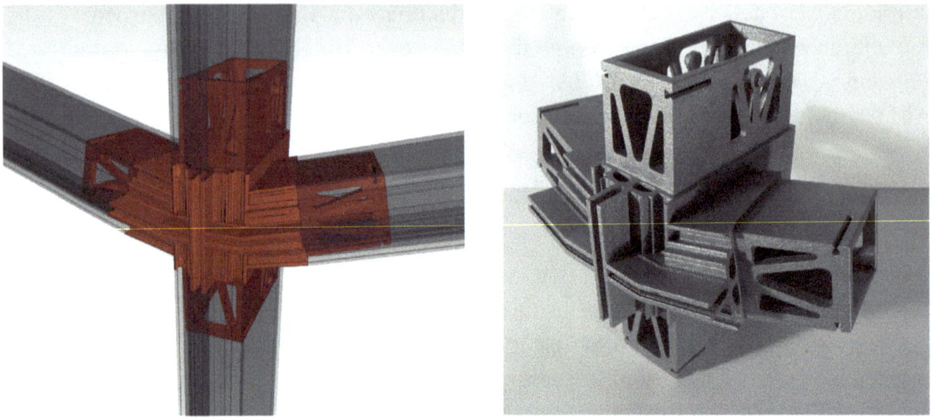

Figure 5.30 Nematox II.

After the presented "Proof of Concept" by Dr. Holger Strauß, only the testing of a real construction project with a larger number of different AM façade parts was still missing.

Observing the latest developments in the building construction process. AM is easy to match/integrate with BIM and digital tools that are applied to the planning and construction of buildings nowadays. AM brings parametrically-controlled complex parts for varying geometries with continuous certification and quality. Parametric design allows for design within material boundaries (e.g., only feasible set-ups supported by CAD-CAM-center).

BIM offers the chance to manage all aspects of informed building construction and to innovatively combine well-known applications. The digitally-optimized façade node also bears the potential to foster a combination of digital planning with digital fabrication into Hybrid Building Construction. Digital design can often neglect on-site construction reality. Well-known materials and products rule the construction process and limit the possibilities at the same time. With the further development of materials and tools/production (e.g., Smart Materials, CNC technologies), the advantage of hybrid building construction comes to the fore.

Now new technologies have to bring the hybrid parts that are needed to combine standard building construction and hybrid building construction. Parts need to offer added value in order to justify their application (cost, liability, risk, etc.).

AM is not the only preferred material/technology, but this dissertation has shown that a new planning process in combination with new AM technologies might be the next step (Strauss, 2013).

5.2.2 N-AM 2nd Generation

N-AM 1st Generation achieved a success by reaching a point that demonstrated the viability of a parametric node. The challenge occurred in the kernel of the node as it could only be accommodated by regular geometrical forms. To increase the possibilities that the node can accommodate, the N-AM 2nd generation was designed. The main goal was to integrate the whole process in one algorithm (script) using Grasshopper and Rhinoceros. The process consists of the generation of the basic node shape based directly on its position in the structure, the structural analysis of the structure, the geometrical optimization of the basic shape of the node through mesh relaxation (minimal surface), and the topology optimization of the nodes based on its stress mesh analysis (Figure 5.31).

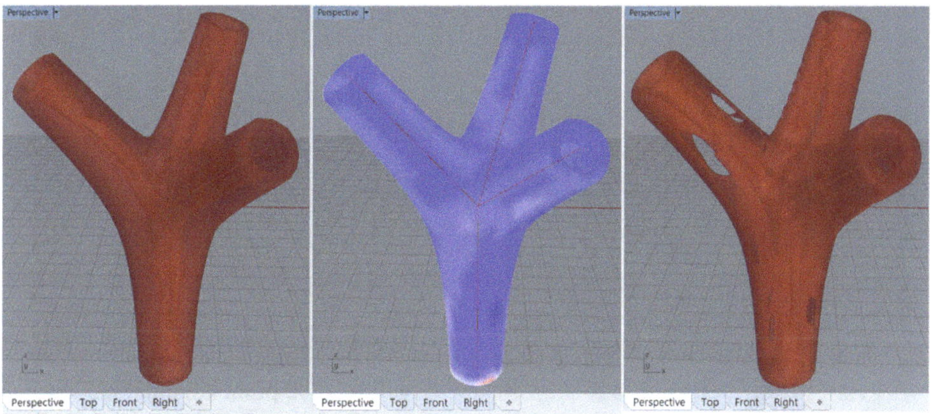

Figure 5.31 mesh relaxation (left), stress analysis (middle), topology optimization (right).

The basic shape of N-AM 10 was four round hollow profiles that connect at a point. The mesh relaxation was done using Kangaroo (Live Physics engine for interactive simulation for Grasshopper). Then the forces were applied on the node, and the stress analysis was done using Karamba (structure analysis plugin for Grasshopper).

The potential of topology optimization has been investigated as a cost reduction factor. There are prominent algorithms in the field of topology optimization, most notably SIMP – Solid Isotropic Material with Penalization. Its stress criterion should be as simple as possible regarding the stiffness-density relation. The 0-1 formulation is a function called density and associated with the design domain of each element in the finite element formulation. The density of zero and one is viewed as solid and void material (Bendsøe & Sigmund, 2003).

The stresses on the node had to be remapped to allow the control reduction of the material. The mesh cells with low-stress values were removed. The form was affected by the cell mesh type.

N-AM 10 was generated as the first physical prototype, and it was printed in Stainless-Steel GP1 using EOSINT M270 system (Figure 5.32). The node was not tested to validate the process. Yet, it proved the potential of integrating the whole process in a one-step operation.

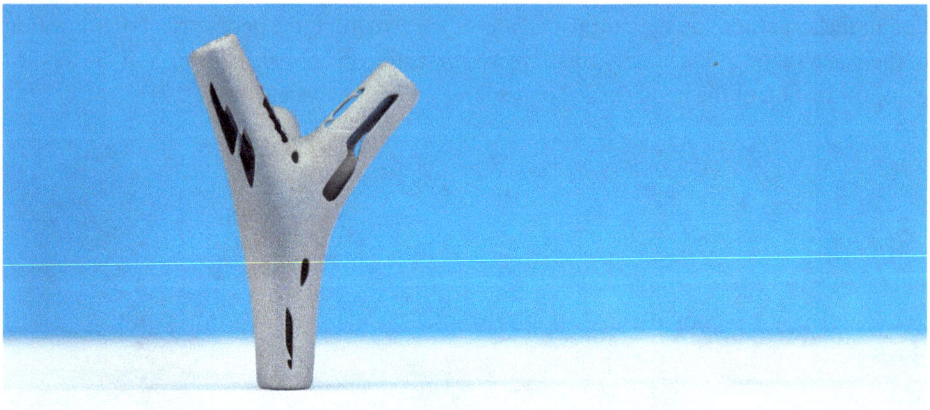

Figure 5.32 N-AM 10.

N-AM 10 was designed to a simple cable structure. The process contained several scripting bugs that were present at the subtracted areas at the node. The generated shape was strongly affected by the meshing process. It was decided not to follow the optimization script for further development, as a suitable software was available later on the course of the dissertation and it was beneficial to utilize it instead.

The next and final stage was the design and engineering of a functional mockup that contains significant required information that can help to obtain a certification for façade/building industry. This information is related to proving the structural integrity of the 3D printed node. N-AM_Li3 presents the first fully functional/structurally tested 3D printed topology optimized façade node for standard façade systems. The testing series and the design process provide all the necessary information to prove the structural integrity of the node. The testing series starts from tensile tests and goes on to component tests and a validated simulation method. The design process starts from the basic geometry and goes to a remodeled and calculable model that is suitable for 3D printing and FE analysis.

5.2.3 N-AM_Li3

The testing of the whole process to design and engineer a façade element was needed to grant the quality of the proposed planning method. N-AM_Li3 was the last project of this dissertation and the goal was to generate a node that can be tested. This is the transition that was needed to validate the topology optimization process and the remodeling of the node. It also served as a second validation for the material model and the simulation process that was proposed as a result in chapter 03.

The façade element is 1.5 * 3.5 m and is formed out of triangles. The focus was given to the nodes testing; accordingly, the design was fixed at the first stage (Figure 5.33). The engineering of the profile was the next stage to define the size of the nodes.

Figure 5.33 Façade mockup for N-AM_Li3 nodes.

The façade element has a rectangular form that contains two internal nodes and six corner nodes. The two internal nodes project outwards from the plane to different distances. The corner nodes are defined as support points, the upper two (1&2) are fixed and the lower (3&4) are movable. All nodes are defined as rigid connection points (Figure 5.34).

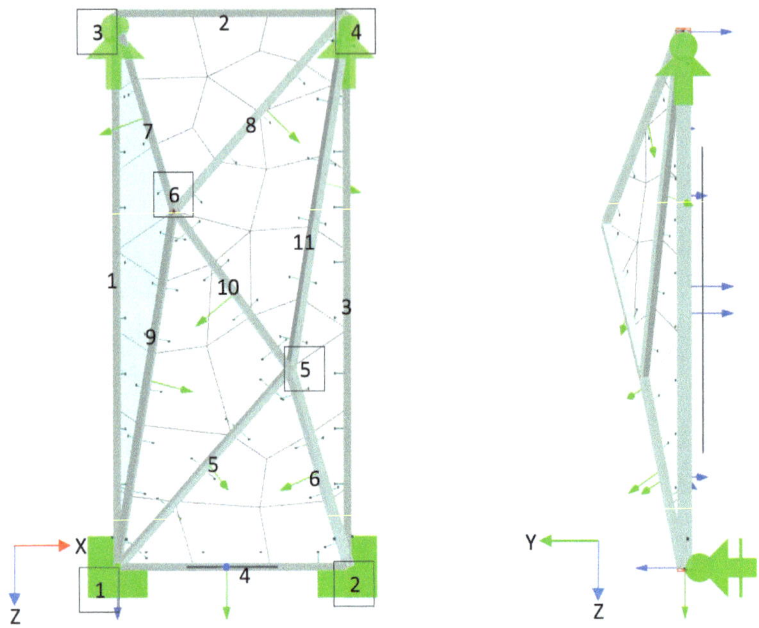

Figure 5.34 The structure system of the façade element.

Aluminum with an isotropic linear-elastic material model was chosen (Table 5.6).

Table 5.6 Aluminum mechanical properties for the calculations.

Material mechanical properties	Abbreviation	Value
Coefficient of elasticity	E	70.000 N/mm2
Shear modulus	G	27.000 N/mm2
Poisson's ratio	v	0.296
Density	γ	27 kg/m3

Glass with an isotropic linear-elastic material model was chosen. The glass was defined as two 8 mm panels (Table 5.7).

Table 5.7 Glass mechanical properties for the calculations.

Material mechanical properties	Abbreviation	Value
Coefficient of elasticity	E	70.000 N/mm2
Shear modulus	G	28455.3 N/mm2
Poisson's ratio	v	0.230
Density	γ	25 kg/m3

Two different cross-sections were selected for the aluminum profiles. There were 50x50 mm for profiles 5, 7, & 9, and 100x100 mm for all the other profiles (Figure 5.35).

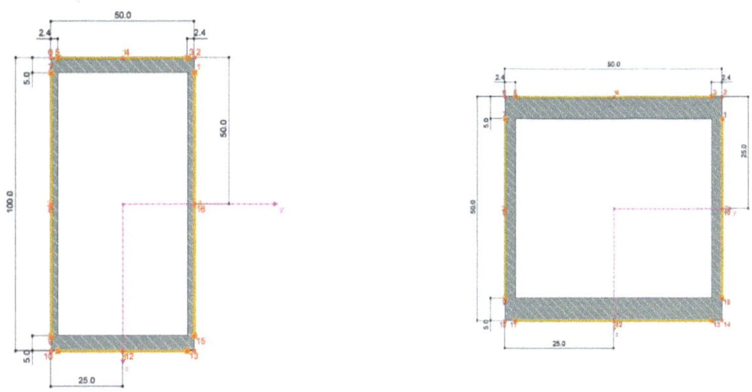

Figure 5.35 The selected profiles 100x100 mm (left), and 50x50 mm (right).

DIN EN 1990:2010-12, DIN EN 1991-1-1, and DIN EN 1999-1-1 were the standards that were used for the calculations. All different load cases such as wind loads (suction and compression) and self-weight were determined and the load combinations were calculated. The stress analysis resulted a maximum von Mises equivalent stress of 2.44 kN/cm2 at node 6 (Figure 5.36).

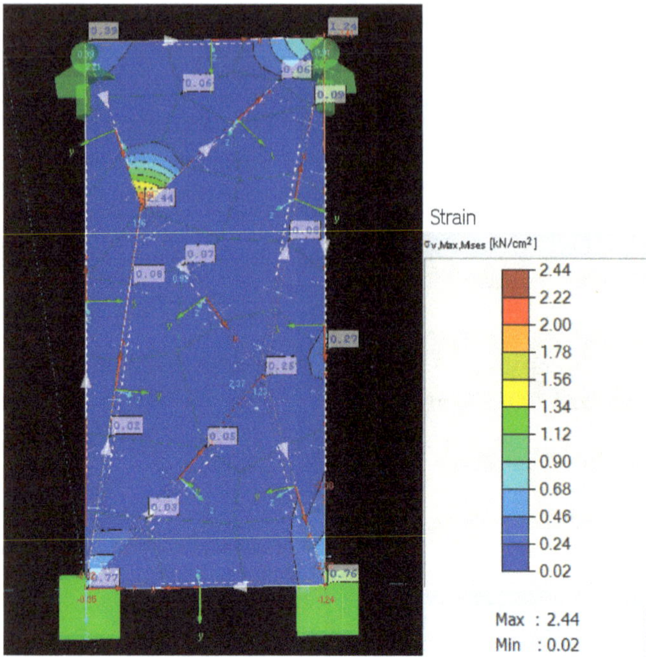

Figure 5.36 Mises stresses of the elements of the façade unit.

Accordingly, node 6 will be designed based on the defined cross-sections. To run the topology optimization, certain material properties had to be defined. In chapter 03, the tests were run in relation to the applied machine (3D printer). In this project, a new 3D printer from the company RENISHAW GmbH and Aluminum (AlSi10Mg) were used. The mechanical properties of the used Aluminum powder (AlSi10Mg-0403) lie within the range of the previously tested specimen in chapter 03 (see Appendix H). Therefore there were no further tensile tests and the material model was furtherly used. It is advisable so to run tensile testing when using a new 3d printing machine.

The node was modeled and prepared for topology optimization. The assembly type had to be defined to generate the geometry of the node. Three different types of connection between the node and the profiles were investigated. The first type (Figure 37-a) was a spigot connection, which is fixed by a screw that passes through the screw channel at the profile and the node insertion. Second and third types were inspired by the dovetail wood connection. The second type (Figure 37-b) is to be assembled from the back side of the node and fixed with two screws at the same side. The third type (Figure 37-c) is to be assembled from the front side of the node, and fixed with two screws that pass through the drainage channel to the node.

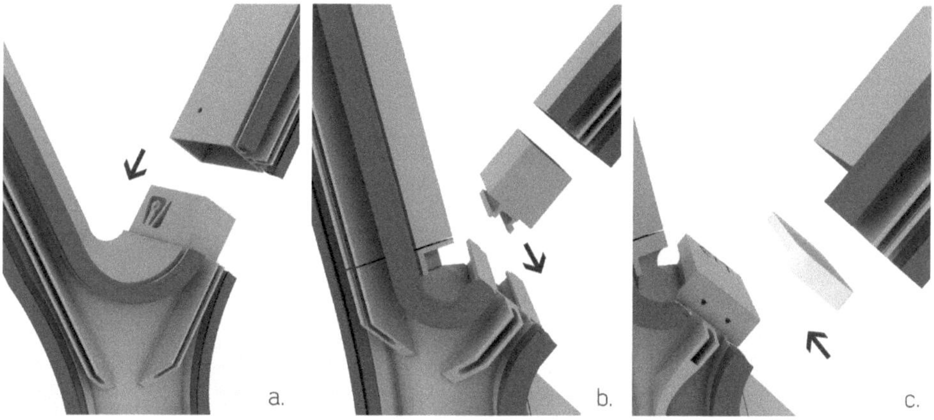

Figure 5.37 The investigated three types of assembly.

The continuity of the drainage channels and the gaskets channels has to be achieved. Thus, the front side of the nodes was a result of profile orientation and the average surfaces that connects these channels.

The topology optimization was done for the first node. The optimization yielded acceptable results (Figure 5.38). Still, the node size was too big for the frequent print volume of the metal 3D printers, and the form was not perfectly designed for additive manufacturing. Accordingly, the idea shifted to use a slide connection to reduce the node size and tighten the curve between the profiles.

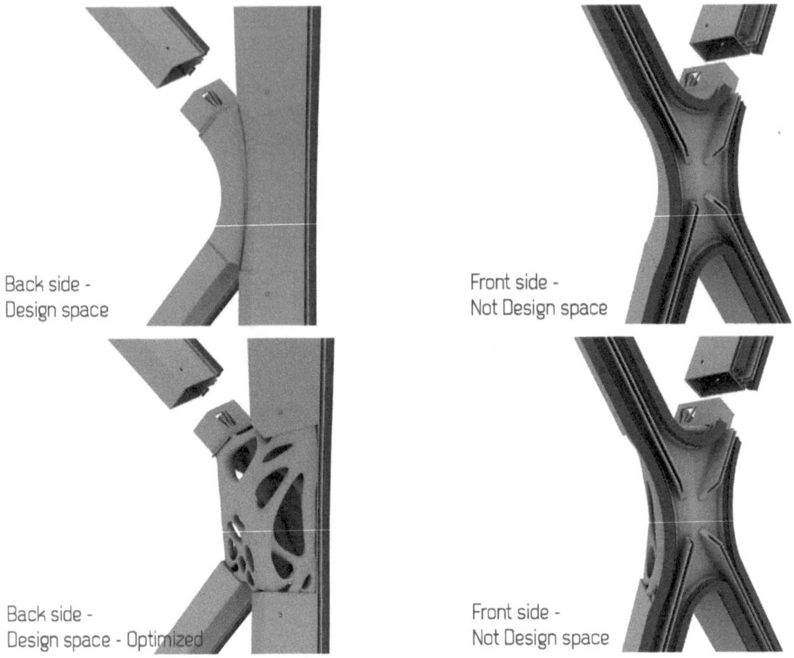

Figure 5.38 N-AM_Li3 – 1st Version.

The second version was relatively smaller, so that two nodes are able to pass into a build plate of 250 x 250 x 300 mm. The sliding connection from the backside helped reduce the size of the node. However, the contact area between the connection and the design space was smaller, which made the optimization denser and built kind of a neck area between the optimization results and the basic not optimized part of the node (Figure 5.39).

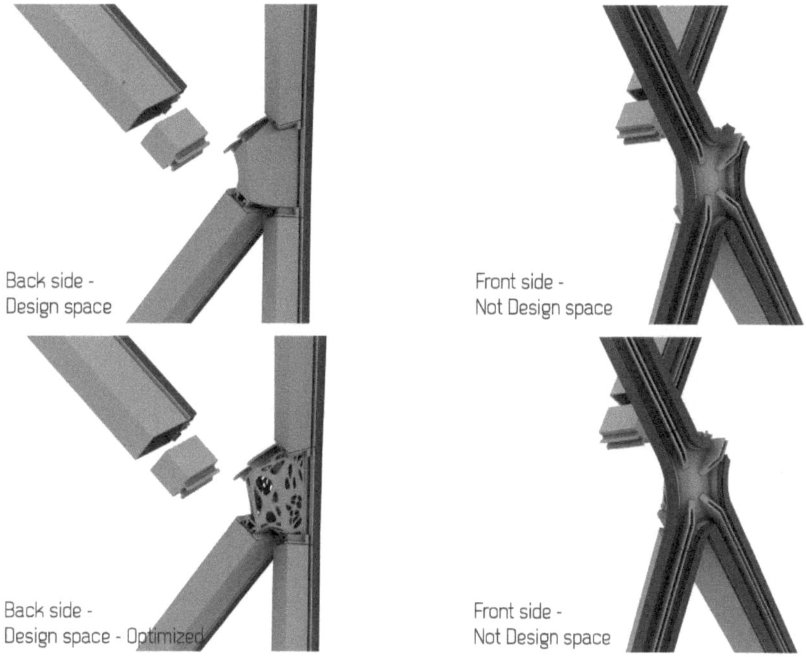

Figure 5.39 N-AM_Li3 – 2nd Version.

The last version of N-AM_Li3 was developed to avoid the previous setbacks. It is slightly smaller than the second version and connects to the profiles through a sliding dovetail connection that is oriented along the profile's normal vector. Two screws that are located at the drainage channels fix the profiles to the node. This version has bigger contact area for the optimized form. The node is designed to have less support structure and it respects additive manufacturing requirements (Figure 5.40).

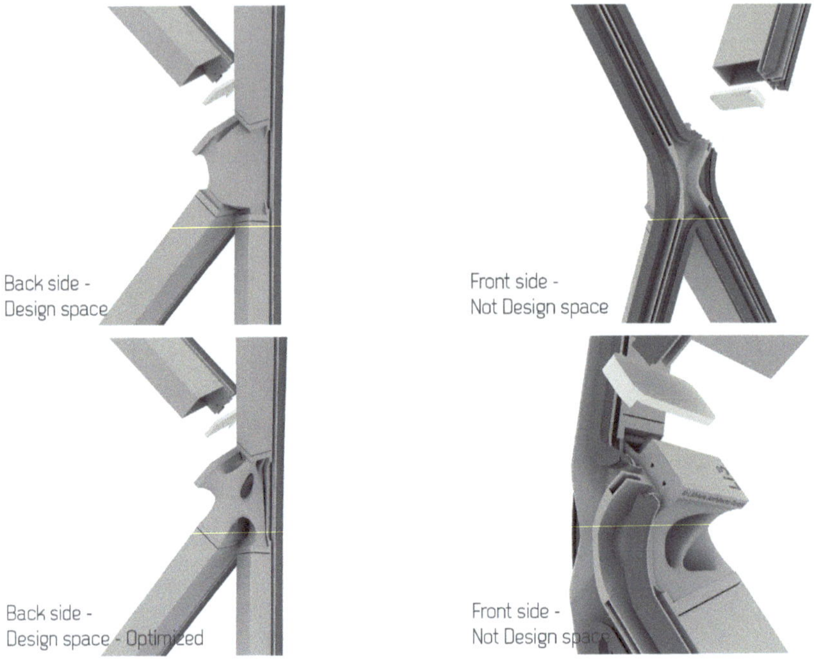

Figure 5.40 N-AM_Li3.

After defining the geometry of the node, the topology optimization was done with the objective to maximize stiffness. The mass target was 30% of the total design space volume with no frequency constraints. All load cases were calculated separately (Figure 5.41).

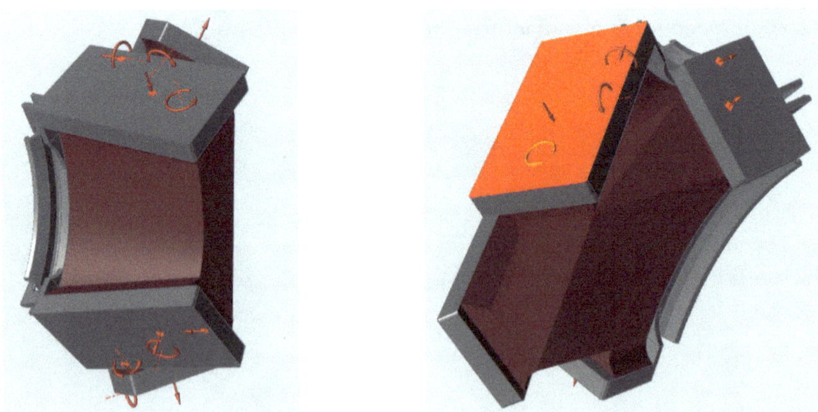

Figure 5.41 The forces that affect the node.

The minimum thickness laid between 8 and 11.23 mm based on the load case. Gravity was considered in all load cases, which means the own weight of the node was included. The different optimization results were overlaid to generate a new boundary that can absorb all different load cases. The orientation of the node in the build volume was defined to help improve the remodeling of the node (Figure 5.42).

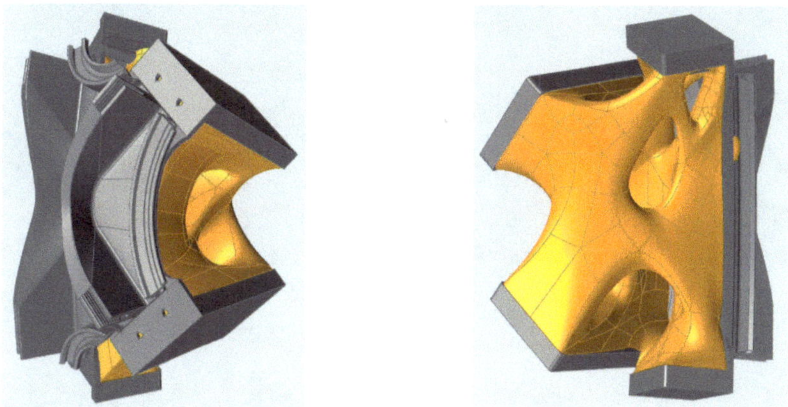

Figure 5.42 The remodeled optimization results.

The final result was one solid volume (Figure 5.43). The synthesis between the design space and the nondesign space is needed to provide a solid volume for 3D printing (Figure 5.42).

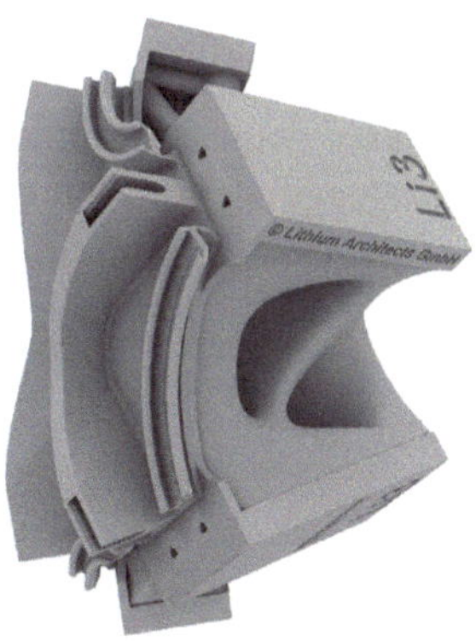

Figure 5.43 illustration of N-AM_Li3. (patent registration number: 10 2019 129 283.2)

This version of N-AM_Li3 was fully developed to move from a prototype to a product. All aspects such as manufacturability, assembly, durability, structural integrity were considered (Figure 5.44). However, all these aspects were investigated in relation to the applied machines and systems that were available during the course of the dissertation. Any change of façade system, 3D printing machines or parameters, and material choice may need further investigation.

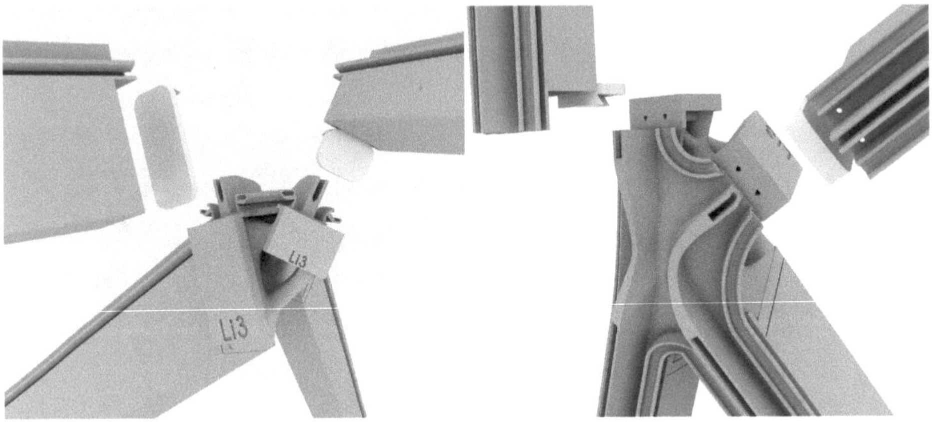

Figure 5.44 N-AM_Li3 and its assembly components.

The last step to validate the structural integrity of the remodeled node and material model one more time (see chapter 03) in relation to the node is the component test. The component test is designed to test the maximum loading capacity that the node can reach before it breaks down. It consisted of the extraction of the node and its connected profiles, then rotated 90° clockwise about the origin of the mockup. Thus, the structure is laid on the ground. The aluminum profiles were replaced with steel profiles to ensure the occurrence of the failure at the node, not the profiles. A cylinder was integrated at the top point of the node and aligned to its normal vector. The cylinder is needed to provide an appropriate contact surface to the testing machine that press the node from the top (Figure 5.45). An FE calculation of the component test was carried out using ANSYS-Workbench to determine the maximum loading capacity of the node and then was compared to the measured forces that came out of the test.

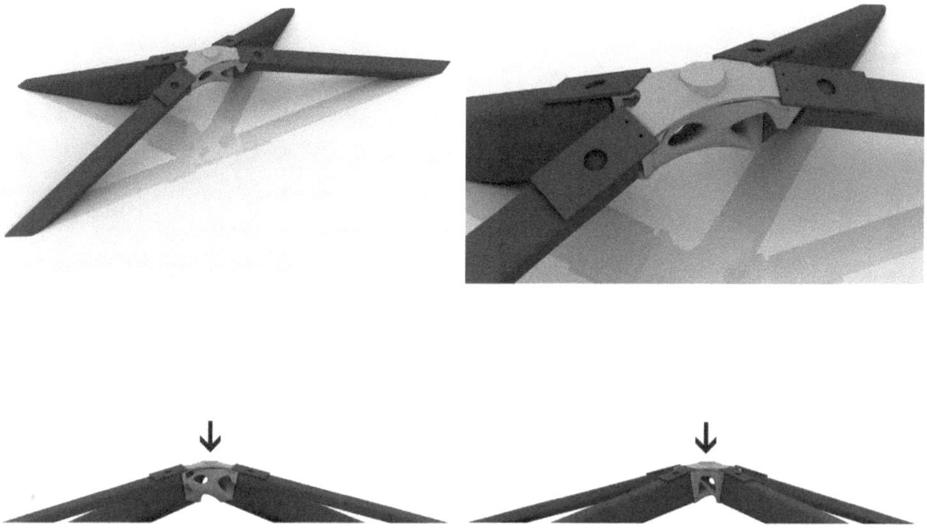

Figure 5.45 N-AM_Li3 component test.

To the run the FE calculation, the model was simplified by removing the connection sliding sides between the node and the profiles. The inner edges of the profiles were defined as support lines. The corners of the support lines were defined as support points to prevent rotation. Figure 5.46-A, the displacement along z-direction for the inner edges was prevented. Figure 5.46-B, the displacement along x- and z-directions for the corner points was prevented. Figure 5.46-C, the displacement along y- and z-directions for the corner points was prevented. Figure 5.46-D, gravitational acceleration to take self-weight into account was considered. Figure 5.46-E, the displacement was caused by the applied force on the cylinder (Figure 5.46). The material model for aluminum that was generated in chapter 03 was used for the calculations, and an isotropic linear-elastic material model was selected. For the profiles, a standard building steel material model from ANSYS-Workbench was used.

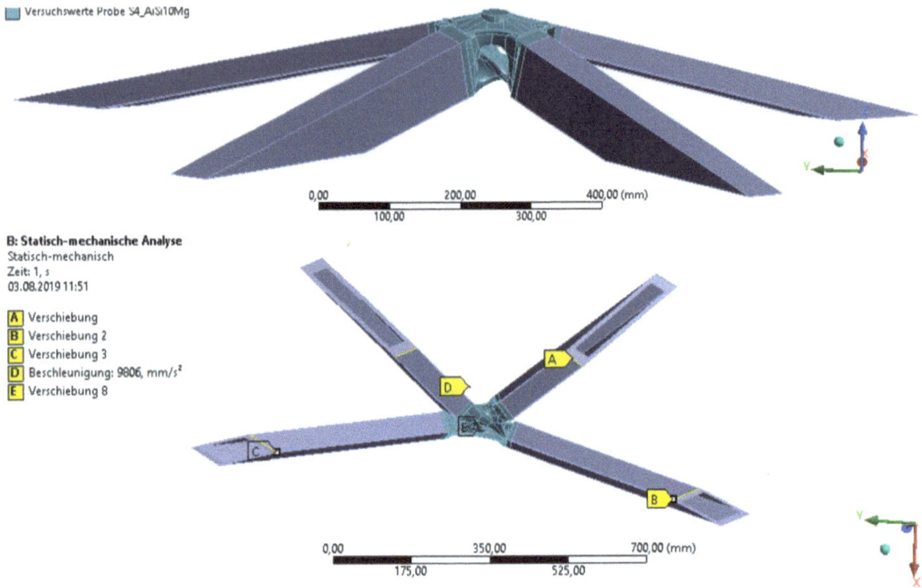

Figure 5.46 N-AM_Li3 component test – ANSYS Model and supports definition.

The meshing of the model was generated. The profiles were meshed using Hexagonal-Dominant method with a roughly 5 mm element size. The node was meshed using prismatic elements with an element size of 2 and 3 mm. The mesh sizes were tested and compared to check the effect of the mesh size on the calculations. The results showed no significant change on the resulted equivalent stress. Accordingly, 3 mm element size was used (Figure 5.47).

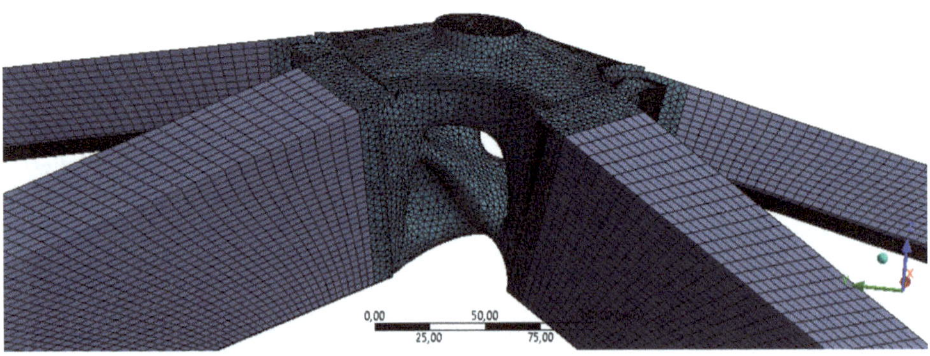

Figure 5.47 FE-mesh of the model. 3 mm element size at the nodes, and 5 mm element size at the profiles.

The next step was the determination of the displacement where the node breaks. The maximum equivalent stress was to occur at the outlines of the cylinder surface. The maximum equivalent stress was searched within the node. Several positions exhibited stresses that have almost the same maximum stress. The displacement was applied linearly and constantly increased over one second. The search was targeting the point in time at which the maximum tensile stress of 289 MPa was exceeded (Figure 5.48).

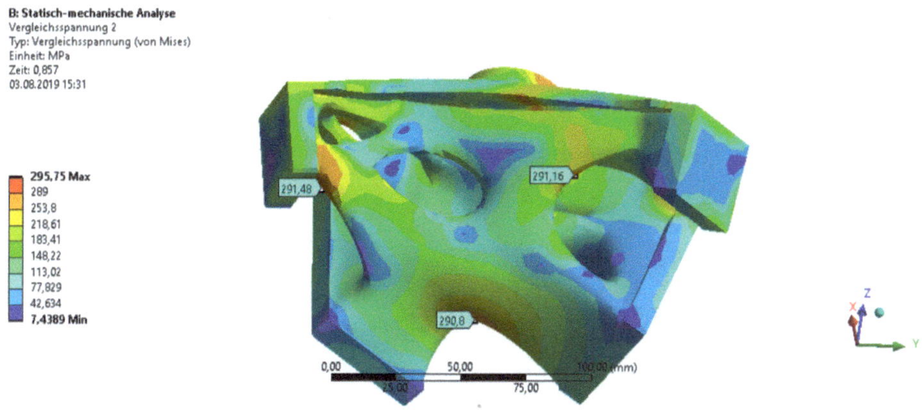

Figure 5.48 Maximum equivalent stress at the node. The marked locations are where the ultimate stress is exceeded.

The force reaction on the cylinder surface at 6 mm displacement is ca. 63.07 kN.

As a final step, the force reaction was applied to the node (force-controlled model), and the results were compared to the displacement-controlled model (Figure 5.49). The stress pattern was very similar, and the equivalent stress values were slightly higher.

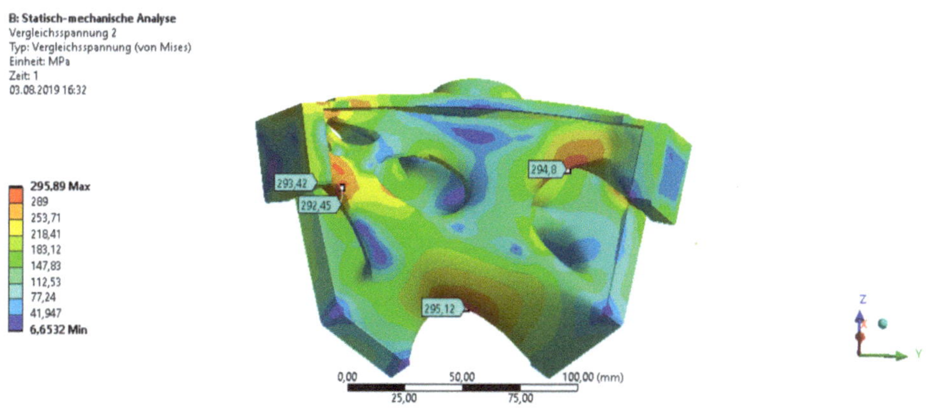

Figure 5.49 Maximum equivalent stress at the node by applying 63.07 kN on the node.

To conclude, the calculations with the program ANSYS-Workbench showed that the node fails at a displacement of ca. 6 mm and a force of ca. 63.07 kN.

The final step was to construct the designed component test. The original dovetail connection was used to connect the node to the steel profiles. They were milled steel parts with the exact original size designed for the mockup. The milled dovetail connections were then welded to the steel profiles. The overlapping at the top of the node was done using 4mm steel plates that were then welded at the top of the steel profiles and connected to the node with two self-tapping (3.5 mm diameter) screws (Figure 5.50&51). According to the dimensioned profiles from the originally designed mockup, there were two types of steel profiles were used (50 x 30 mm, and 90 x 50 mm).

Figure 5.50 N-AM Li3 component test - physical mockup.

Figure 5.51 N-AM_Li3 component test – Steel construction connected to the N-AM_Li3 node.

To mimic the support conditions as the ANSYS model, a ball connection was used to support the legs of the component test (Figure 5.52).

Figure 5.52 Support design as built for the component test.

The components test was then started and pressed on the node with 4 mm per minute. The breaking point occurred at the dovetail connection that was neglected at the simulation at 10 k.

6 Li3-Method

The first draft of the planning concept that is mentioned in 1.3 consisted of three stages: digital conception, manufacturing conception, and assembly process. Throughout the course of the dissertation, it was beneficial to merge the assembly process with its logistics in both digital conception and manufacturing conception by means of designing elemental structures that respect most commonly transportable sizes. The logistics topic itself is related to the country where a project will be built. Thus, to widen the applicability of the method, this topic is left to the project manager to investigate. Li3-Method is the new planning concept with which this dissertation concludes (Figure 6.1).

Li3-Method presents a planning concept to generate complex façade and building envelopes that can be constructed by applying standard mullion-transom façade systems. The mullion-transom system combined with the N-AM_Li3 node (patent registration number: 10 2019 219 283.2) can be constructed as a stick system, unitized system, or free-form shell structure, in either steel or aluminum.

Li3-Method represents different stages and sub-stages that complement each other. The different stages are classified under four main phases. These phases are Global Geometric/Structural Phase (blue), Local Geometric/Structural Phase (red), Validation-of-Manufacturing Phase (green), and Façade Erection Phase (orange) (Figure 6.1).

The Global Geometric/Structural Phase is generated based on the collected knowledge presented in chapter 04. In chapter 04, the basic understanding of the different scientific approaches to generate an elemental structural model in a digital medium were investigated. The rules of free-form façades/building envelopes and their engineering were explained. The possible optimization methods on the global level such as Evolutionary algorithms and Multi-objective optimization were described and tested in chapter 05.

At this phase, the projects evolve from an idea into constructible components. The optimization objectives vary between geometrical and structural objectives. There are endless possibilities to combine these objectives. In Voronoi-façade, there was one structural optimization objective. This objective was to find the lightest weight of a structure by moving its control points within a range of coordinates. In Paneling-Free-Form, there was one geometrical optimization objective. This objective was the planarity of the glass surfaces. In Global Structural optimization of Grid Voronoi shell structure and Topology optimization of its nodes, there were three optimization objectives. These objectives were reached through two optimization processes: one objective geometrical optimization (regular cell distribution) and two structural optimization objectives (less mass and less displacement). The table, with one optimization objective (minimum displacement),

A. Mohsen, *Design to Manufacture of Complex Building Envelopes*,
Mechanik, Werkstoffe und Konstruktion im Bauwesen 56,
https://doi.org/10.1007/978-3-658-30204-7_6

aimed to investigate the remodeling of the raw topology optimization to improve the 3D printing process and reduce support structure. In Façade Technology 1, the attention was made to control the diversity of the manufactured elements by running one objective geometrical optimization to equalize the profiles' lengths.

The final design is defined by the end of this phase. To be able to proceed with the local geometric/structural phase, the validation of the manufacturing process phase has to be executed.

The validation of the manufacturing phase is needed because of the integration of additive manufacturing of metals in the planning process. Additive manufacturing is not regulated by the building industry. Therefore, a sufficient set of information about the structural behavior of the 3D printed metal elements and a method to predict its behavior is needed. The set of information is represented by the mechanical properties of the 3D printed elements that were discussed in chapter 03. In chapter 03, tensile tests were done on two different metal alloys (Stainless-steel 316L (1.4404), and Aluminum ALSi10 Mg). Tensile specimen Form A mentioned in DIN 50125 was used for the tests. The results were the mechanical properties of two print direction of the specimens (vertically 0°, and tilted 45°). The weakest were used to construct a material model in ANSYS. This material model was used to design and engineer a validation model. The designed validation model is a form that when put under tension, a complex load condition occurs in the predefined location not only tension force as in the tensile specimens. The material model was designed to behave isotropically. The goal was to simplify the simulation process of the 3D printed elements, as the modeling of micrometer layers of the original part is complicated and needs enormous computing power.

The validation models were printed in four orientations: horizontal 90°(L), vertical 0°(S), flat 180°(F), and tilted 45°(45). The goal was to compare the simulation results with the actual results of these main four print orientations. The behavior of the simulated model was slightly weaker than all printed orientations, which proved the made assumption of simulating 3D printed metal elements as casted parts. Additionally, for structural safety, any designed construction element is not 100% utilized and a safety factor is always considered for the calculations.

Executing these testing stages is recommended for each 3D printer and material alloy that is intended for use. This accurately defines the material model that is needed for the simulation and the topology optimization later at the local geometric/structural phase. The reason for this is that each manufacturer prefers slightly different printing parameters, and the used material alloy/its production method affect the mechanical properties of the 3D printed element. However, the validation models could be skipped when the results of the tensile tests of the same material alloy manufactured with two different 3D printers are similar.

The behavior of the used material and 3D printer is now defined. This is considered the first level of validation, which defines the material behavior and all the necessary mechanical properties required for the upcoming stages. To be able to do the second validation level, which is the testing of the node as a component test, the local geometric/structural phase has to be executed.

The local geometric/structural phase is mainly concerned with the design and engineering of the connecting nodes. There are three challenges at this phase: the geometric challenge that is presented as the continuity of the functional channels through the profiles, the assembly challenge that is presented as the type of connection between the node and the profiles, and the structural challenge that is presented as the topology optimization of the node and generating fully functional/tested topology optimization node.

The investigated geometric rules, design for manufacturability and assembly, and topology optimization in chapter 04 helped to refine the node generation through the projects presented in chapter 05. The development of the node generation started in Highrise-tessellation when a simple node between round profiles was parametrically generated. The focus on using standard mullion-transom systems made the requisite in generating a node for rectangular profiles that could be parametrically generated based on a grid of lines and cross-section measurements. This was achieved in Voronoi-façade. A fully parametric node for rectangular profiles was designed. The node was tested for more geometrical conditions to stabilize and debug the script in the project Paneling-Free-Form. The final geometrical development was presented in N-AM_Li3 node.

N-AM_Li3 presents a fully functional geometric solution to connect standard mullion-transom systems. The outer side of the node grants the continuity of the drainage, the screw, and the gasket channels. The curved transition between the profiles grants the tightness of the connection between the glass panel and the system with constant gasket size. The backside of the node accommodates the irregularities that are generated by the different profile inclination. The connection between the node and the profiles was developed from a sliding connection that is parallel to the profile axis via a sliding dovetail like connection. Two assembly scenarios of dovetail like connection were investigated. These two scenarios were when the profiles are assembled from the backside of the node, and when the profiles are assembled from the frontside of the node. The last lead was the best solution, and an overlapping connection was designed. In this connection, the upper side of the profile rests on the sliding side of the node and is then fixed with two screws. The profiles are cut in an L form to achieve an appropriate connection. Additionally, the opposite side of the sliding connection is to be integrated into the profile's cavity.

Topology optimization was investigated first in N-AM 2nd generation when a simple cable connection was designed. The form-finding was done based on a minimal surface rule and the weight reduction was based on the less stress mesh cells within a predefined mesh. The generated model was checked and presented the ability to absorb the subjected tension loads. The process was utterly self-programmed.

The development of trusted topology optimization software and the further development of a node for rectangular profiles presented certain geometric conditions. The node consists of two parts but only on the digital level, design-space and non-design-space. The design space is the main geometry of the node that is responsible of the functional continuity through the structure. The non-design-space is the volume that accommodates the geometrical irregularities between the profiles. This digital generation concept is needed to run a successful topology optimization. The generation of these elements was parametrically designed, and the topology optimization was tested and investigated in global structural optimization of Grid Voronoi shell structure and topology optimization of its node. In this project, the generated parametric nodes for rectangular profiles were used and the concept of defining design-space and non-design-space was considered.

In the same project, different percentages of weight reduction were tested (30%, 20%, and 15%). Topology optimization is an iterative process and it was needed to define the most appropriate weight reduction percentage and that was 30%. Two types of connection (welded and bolted) between the node and the profiles were tested to determine the connection type effect on the topology optimization results. It was determined that the connection type affects the topology optimization results. Still, the topology optimization was for separate load cases, the material mechanical properties were standard construction steel, and the raw optimization presented rough surfaces based on the mesh size used. These challenges were solved in Façade Technology 1, when all load cases were optimized separately, then overlaid, and a new model was generated to contain all generated volumes. The new model is expected to withstand all calculated load cases and provides a smooth surface that is better for 3D printing. Still the generated node was not analyzed to ensure that the new geometry can absorb all subjected loads, and the material mechanical properties were standard construction steel.

In N-AM_Li3, a life-size façade unit was designed (1.5 x 3.5 m). The stress analysis according to DIN EN 1990:2010-12, DIN EN 1991-1-1, and DIN EN 1999-1-1 was calculated. The profiles sizes were determined and the most stressed node was defined. The topology optimization with 30% weight reduction was calculated for all load conditions. The Aluminum material model for topology optimization was the generated material model in chapter 03. This guaranteed the accuracy of the results. The generated node N-AM_Li3 fulfills all façade/structural functions of the façade element. Yet, a component test has to be executed. This node test is the last stage in the validation of manufacturing phase, and it is the second level of validation that will validate the generated material model for the second time, and the generated topology optimization remodeled geometry for the first time. This permits a physical test of only the most stressed node of the mock-up and a digital-only check of the other nodes.

At this stage, a component test was designed at N-AM_Li3. The goal was to end the validation of manufacturing phase. The highest stressed node was defined and engineered. The component test consisted of a steel structure that was connected to the node and was subject to compressive loads until failure. The node is manufactured out of Aluminum AlSi10 Mg and 3D printed using a 3D metal printer from the company RENISHAW.

The results of the maximum force that the node can take and the failure position was compared to analyzed/calculated component test in ANSYS using the material properties of Aluminum AlSi10 Mg that was defined in chapter 03. Therefore, the process moved back to the last stage at the local geometric/structural phase. This stage was to analyze the node to predict the results of the component test. The analysis started by determining the maximum displacement of the node before it broke. The reaction force that was generated at the breaking displacement was again subject to the node as a compressive force. The results were then compared and in N-AM_Li3, the two simulation results were almost identical. These results are then compared to the physical component test. In N-AM_Li3 mockup, the component test was done but the breaking point occurs at the dovetail connection that was neglected at the simulation under 10 kN compressive force.

All these calculations were collected in the form of static reports that include all the exact calculations of the elements. The FE elements model is precisely defined and all the calculations settings are summarized in an FE report. All required information that is needed for manufacturing is contained in the parametric model, such as the length of the profiles, the size of glass panels, the digital nodes geometry, and the reaction forces inside the mock-up. All the information can be extracted in the form of tables if needed. This information represents the final data set of the structure and the beginning of the façade erection phase. At this stage, the set of data could be used to obtain a certification of using the 3D printed nodes within the structure. The logistics are left to the project manager, putting into account that the topic logistics were only considered in terms of designing transportable elemental structure.

The mechanical properties of 3D printed parts may differ with the change of the processing parameters, such as laser power, hatch distance, and metal powder morphology. The processing parameters usually vary from manufacturer to the other. Each searches for the best settings that suit his machines and materials.

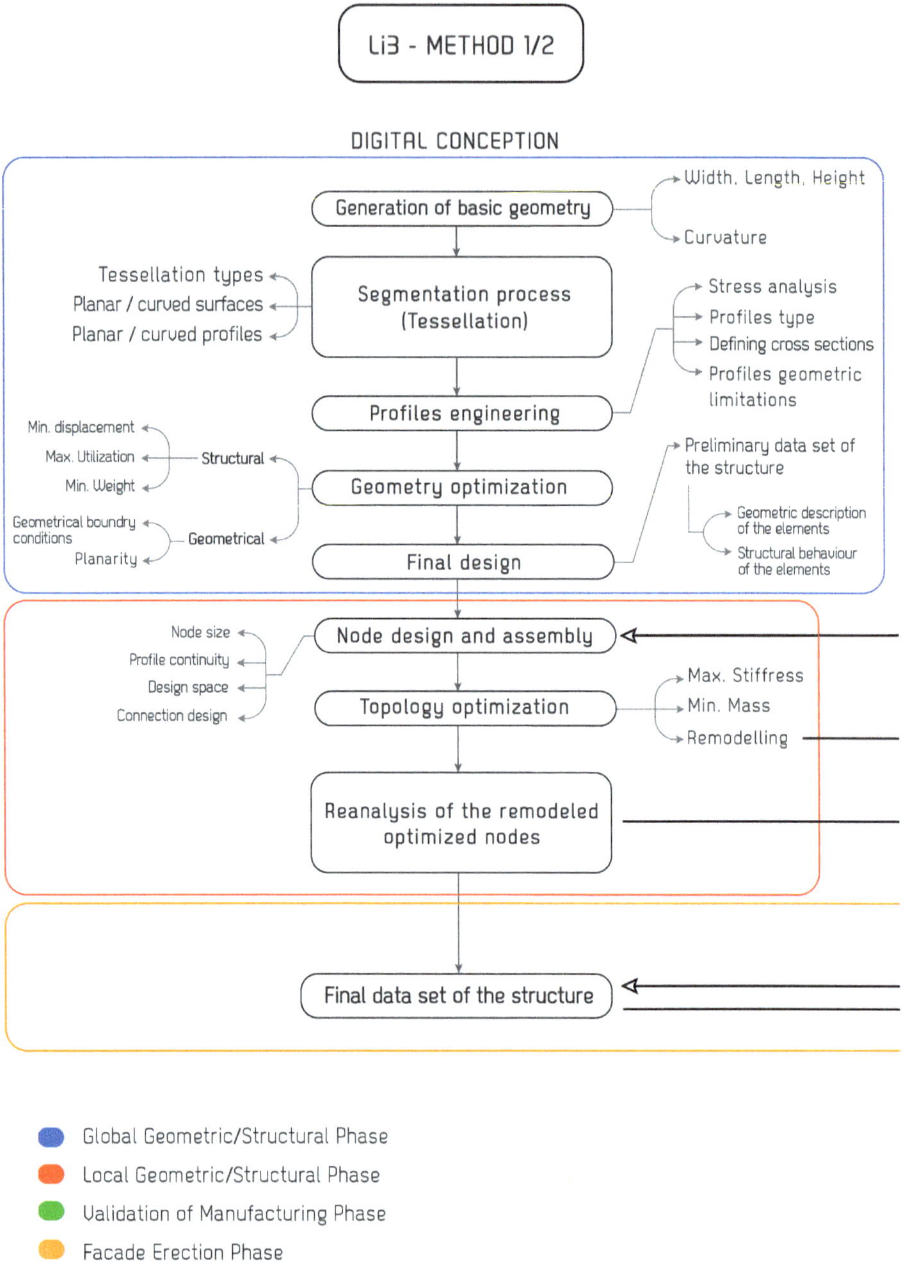

Figure 6.1 Li3-Method 1/2.

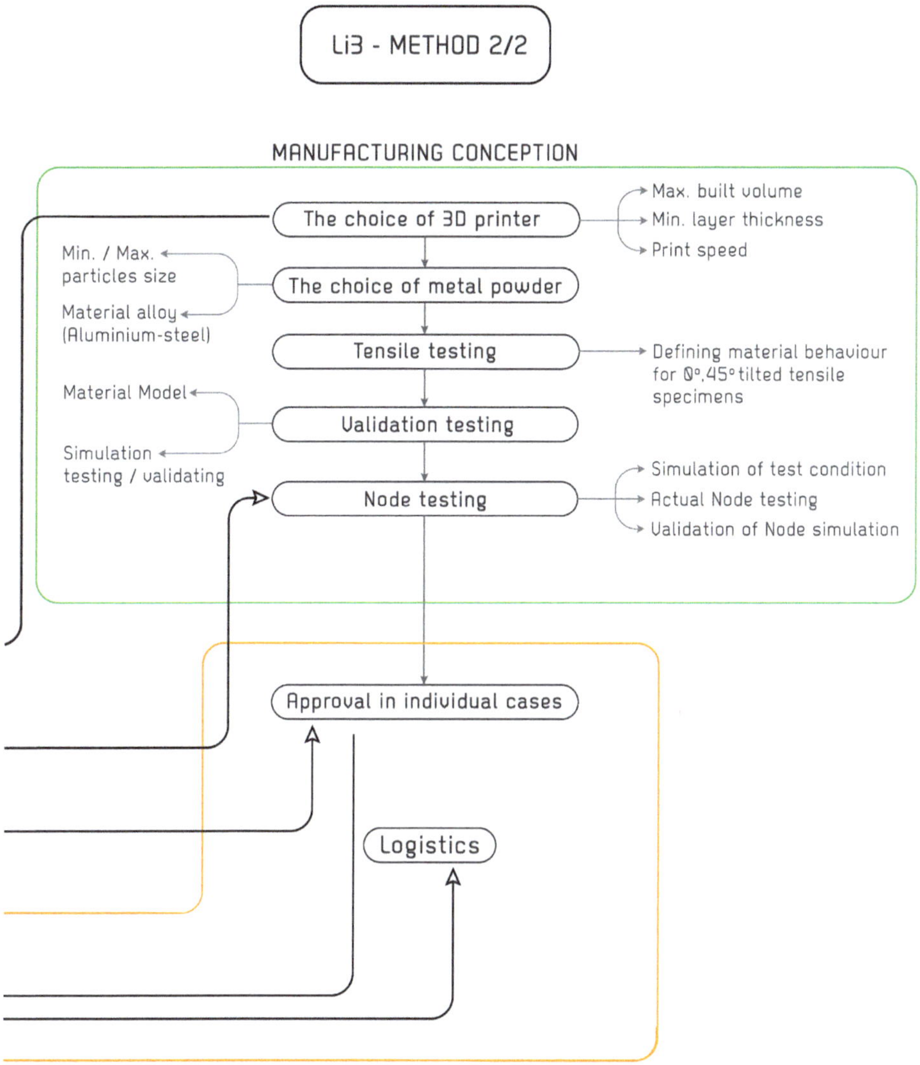

Figure 6.2 Li3-Method 2/2.

Li3-Method consists of two main stages: digital conception and manufacturing conception. The two stages run in parallel until certain sub-stages coil (Figure 6.1&2).

Digital conception starts with the generation of basic geometry, which defines both the overall dimension of the structure and the curvature of the free-form surface. The segmentation process (tessellation) defines the structural network that represents the free-form surface. This network describes the planar/curved profiles and the planar/curved surfaces that form the structure. At the profiles engineering sub-stage, the structural elements are analyzed to determine the profiles' type, its cross-sections, and its limitations. The global optimization of the whole structure is specified in the geometry optimization sub-stage. This sub-stage is defined by two different optimization objectives: structural and geometrical. The relation between different structural and geometrical optimization objectives can be formulated. The final design is now generated, and it represents the preliminary data set of the structure. The global geometric/structural phase is then complete.

Manufacturing conception starts with the choice of the 3D metal printer. The maximum build volume, minimum layer thickness, and print speed now are established. The metal powder is selected based on several aspects such as min./max. particle size and material alloy (aluminum and steel). The tensile testing defines the material behavior by testing tensile specimens that are tilted at 0°, and 45° to the Z-axis of the build plate of the 3D printer. The validation testing uses the results of the tensile testing to construct a material model that is used to run the validation testing. The simulation of a validation model is done to compare the behavior of different print orientation of a complex element to a simulation model that is treated as a casted part. This validation allows ignoring the print direction of the elements, which simplifies the simulation process. At this sub-stage, the geometry of the 3D printed parts must be defined in order to start the sub-stage node testing and finalize the validation-of-manufacturing phase.

Back to the digital conception, the node design, and assembly can start. According to the engineered profiles and the maximum build volume of the chosen 3D printer, the size of the nodes is determined. The nodes have to ensure the continuity of the profiles and their drainage, an appropriate design space for the topology optimization, and an engineered connection design to the profiles. The topology optimization is done by applying the forces that are generated in the preliminary data set of the structure. The generated optimized nodes are then remodeled with respect for design for manufacturability and assembly guidelines, to enhance the 3D printing process. The remodeled nodes are moved to the sub-stage node testing in the manufacturing conception.

At the node testing, the highest load condition of the nodes is selected based on the stress analysis. The simulation is done for the node, and the actual node is tested. The results of the node test and the simulation are compared to certify the simulation process for all other nodes.

The re-analysis of the remodeled optimized nodes and the node test ends both local geometric/structural phase and validation of manufacturing phase. The generated final data set of the structure contains valuable information that describes the parts of the element precisely, the mechanical properties of the 3D printed nodes that prove its structural integrity, and a validation simulation method. This information could be used to obtain a certification (approval in the individual case) for the 3D printed nodes. However, other complementary tests may be needed.

The final data set of the structure with its approval in the individual case and its logistics devise the façade erection phase.

Li3-Method considers the digital planning and generation of construction information, but not the tendering and commissioning of the different building stages, which are affected by the new planning sequence.

7 Summary and Outlook

7.1 Summary

The current architectural era demands more transparent structures and highlights single layer envelope structures. These structures have to fulfill their structural and geometrical requirements. Therefore, many node connections have been developed. The node connection is an essential part of the whole system, as it has to bear not only axial forces but also bending moments. The main components of single layer envelope structures are connecting bars, planar surfaces, and connecting nodes. This dissertation was dedicated to the design and engineering process of complex single layer envelope structures, through the exploration of the potentials of the file-to-manufactory approach. This approach is derived from the opportunities of parametric modeling and additive manufacturing.

Drawings express architectural designs. The drawing medium is the primary constraint for drawing acts and organizing ideas, spaces, and functions. CAD software introduces a new capacity for the drawing medium, which is the direct translation of drawings within the digital realm. Those drawings are a digital representation (code) of the main ideas. The code creates geometry as a result, and this is direct modeling, which represents the additive logic of traditional drawings.

Luigi Moretti firstly introduced parametric modeling in 1939. His models of a soccer stadium, where the calculated pseudo isocurves generated a final shape that optimized views from every position in the stadium, were the first representations of associative logic and form-finding (Mulazzani, 2000).

Direct modeling is the traditional modeling technique, where any additional modifications alter the current design spectrum/drawing space. Parametric modeling is driven by parameters, as parameters modifications are reflected directly on the design spectrum/drawing space (Tedeschi, 2014).

Additive manufacturing or 3D printing is one of the computer-aided production techniques that has started to grow in the last few years. It offers new solutions to several problems in the manufacturing world. Additive manufacturing is based on the principle that any object can be synthesized via a breakdown into multiple layers. These layers need to be laid on top of each other to form the final object. Metal-based additive manufacturing has seen significant growth recently because it offers many advantages over traditional

A. Mohsen, *Design to Manufacture of Complex Building Envelopes*, Mechanik, Werkstoffe und Konstruktion im Bauwesen 56,
https://doi.org/10.1007/978-3-658-30204-7_7

manufacturing techniques. Terry Wholer, when commenting on the formnext fair held in November 2016, said "It was the most elaborate and striking display of AM products and services ever." (Wohlers, 2016).

The sequence for the 3D printing process is strikingly similar for all types of 3D printing. This sequence starts with the creation of the digital model for the requested part and then transforms it to a .stl format (stereolithography), which is the suitable format for 3D printing. The generation of the g-code needed to run the printer is then generated, and the print process starts. Finally, post-processing is sometimes needed to improve the object's appearance or enhance its properties. This can include coating, removing support structures, sanding, or thermal treatments.

Single layered structures present different planning challenges. These challenges lay in its structure system. The type of the structural elements and whether it is a standardized system or a customized one affect the complexity of the planning process dramatically.

The tendency in architectural design and façade technology is to increase complexity. Therefore, these types of projects must be approached differently. Baccarini stated in his review in the International Journal of Project Management, "Construction projects are invariably complex, and since World War II have become progressively more so." (Baccarini, 1996). Complex projects demand special treatment, therefore traditional project management is found inadequate. Hough claimed, "complex projects demand an exceptional level of management, and that the application of conventional systems developed for ordinary projects have been found to be inappropriate for complex projects." (Morris & Hough, 1987).

Project complexity has no precise definition. Different papers have discussed the definition of project complexity. It is concluded that project complexity defined by two dimensions, each with two sub-dimensions. The first dimension is structural uncertainty, which is defined by the number of elements and the interdependence of elements. The second dimension is other uncertainty, which is defined by the uncertainty in goals and the uncertainty in methods. Uncertainty in traditional projects where background knowledge is available is less risky than uncertainty in state-of-the-art projects where lack of knowledge is present. Therefore, classical project management techniques cannot handle that level of uncertainty. New methods and ways are needed to design and manufacture modern and sophisticated projects (Williams, 1999).

Standard systems have an excellent Price-Value relationship. However, they are only applicable within specific ranges and limitations. The merge between parametric design, additive manufacturing, value engineering, and standard façade systems will present a planning method that supports the design and engineering of economic complex facades/ buildings envelopes.

The planning concept is based on file-to-manufactory approach. The concept will merge parametric design and additive manufacturing into a method that serves to solve most of the structural design problems for complex building/façade envelopes that are intended to be constructed using standard façade systems, and 3D printed metal connecting nodes.

Selective laser melting (SLM) is one of the additive manufacturing techniques for metals. SLM is an efficient technology to manufacture metal parts made out of different alloys, such as titanium, aluminum, and steel. This efficiency comes from the additive manufacturing method, which means that there is no waste material in the process. It is preferred for light and grid constructions. SLM still faces many challenges. One of these challenges is the physical interactions that happen during the printing process. These interactions develop nonuniform phenomena, which occur because of the high number of control parameters. Thermal gradient and thermal stresses are the primary interactions that can develop in the produced element and may affect its properties. The different sizes of the powder particles and the slow construction rate of the process, as well as the lack of material alloy varieties that are suitable for the process, increase the surface roughness of the 3D printed objects. Accordingly, post-processing may be needed based on the end-use of the object (Gibson, Rosen, & Stucker, 2010).

Parametric modeling increases the complexity of geometries. However, it makes individualized solutions more reliable. For standard manufacturing techniques, individualized solutions are costly and not preferred. On the contrary, AM is not affected by the individualization of solutions. The design and engineering of complex façades and envelopes lead to individualized solutions most of the time. The challenge was the certification of the process and produced parts for the façade industry. In the building industry, additively manufactured parts are not regulated. Accordingly, they are only applied by approval in individual cases that certify the parts and complete the planning concept.

The typical tensile specimens are usually utilized to predict material behavior in the form of ultimate tensile strength, breaking strength, maximum elongation, and reduction in area. The dilemma when it comes to additive manufacturing is the anisotropy behavior, which leads to running the tensile tests for specimens with different print orientations to reach an approximation of the tensile strength. The lowest value is used to create the material model in ANSYS. This material model presents material behavior under stress.

Structure elements are not only under tension forces. Accordingly, a validation model had to be designed and engineered. The goal was to design a complicated loading case that can mimic the behavior of structural elements. The results were used to improve the simulation process of the AM elements and to predict the behavior of additively manufactured construction components. The process is a comparison of the results for the simulated validation model and the tested models, using the material model generated from the tensile tests. The idea is to treat (or test) additive manufactured elements as casted parts, and to reach a value that can be used later during design processes, such as topology optimization, without the need to physically test each element.

Parametric design helps generate complex free-form surfaces using different software and tools. These software and tools have been developing rapidly and are accessible to almost every designer. Thus, free-form designs have become more frequent. Free-form envelopes (in this dissertation) are one layered construction, individual geometric and structural conditions, and developable surfaces. Therefore, the structure is constructed from straight bars that are fixed at each end with a force-fit node. Following economic aspects, the use of planar surfaces that are fixed to the straight bars is considered.

Once the basic (reference) surface is defined, the geometric analysis process starts. Through the analysis, the required tessellations are generated. Tessellations are the replication of the curvilinear architectural forms in symmetrical and interlocking profiles. The development of digital tools and the tendency to construct organic structures allow imitation of biological processes to generate new and complex shapes. The idea behind is to look for originality and to focus on an effective use of resources (Gawell, Rokicki, & Nowak, 2014).

These new and complex shapes are challenging when they are transformed into a system of interlocking profiles. The challenge shows in the connecting nodes between the profiles. The generation process of these nodes can be parametrically defined and controlled. Certain factors affect the generation process, such as the bars' orientation, the node size, and the joinery type. It was essential to consider the façade functions and its influence on the design of the nodes and the whole structure. Façade functions are classified and described in many types of research. Façades primary function is protection against external influences. The secondary functions are user-comfort and economics, and the subordinate functions are aesthetic and structural design (Beyer & Kuba, 2006). The generated structural elements, such as profiles, nodes, and filling surfaces have to fulfill the requirements to achieve the intended façade functions.

The parametric generation of the structural elements for façades/building envelopes allows the performance of optimization processes on the model at the global and local level. It is possible to form some of the intended functions in the form of optimization objectives. To construct a shell structure model, a search range (free-form reference surface) had to be defined. To solve these structures, there are two different approaches: the top-down approach, where the complete structure is broken down into sub-systems and separate tasks, and the bottom-up approach, where the process starts with the individual elements.

Once the model is defined, the global optimization process can be determined. The selected type of optimization algorithm is based on the number of optimization objectives. Some of the most common algorithms to solve structure problems are evolutionary algorithms and multi-objective optimization. Evolutionary algorithms can find possible solutions for particular problems and generate them artificially. Multi-objective optimization (Pareto optimization) is applied when the optimization objectives are fundamentally independent of each other. It generates Pareto optimum, which is a state in which it is impossible to improve one objective without deteriorating the other.

Next, an optimization process on the local level was performed. Topology optimization is used to enhance the performance of the nodes, and to reduce manufacturing costs. In the field of topology optimization, there are prominent algorithms, such as Solid Isotropic Material with Penalization (SIMP). Its stress criterion, such as the stiffness density relation should be as simple as possible. The 0-1 formulation is a function called density, and it is associated with the design domain on each element in the finite element formulation. The density of 0 and 1 is viewed as solid and void material. This approach is used to maximize stiffness and minimize mass of structures through penalization (Bendsøe & Sigmund, 2003).

The generated topology optimized nodes have to be remodeled to generate smooth surfaces that are suitable for additive manufacturing. The remodeling process is controlled by the rules of design for manufacturing and assembly (DFMA). DFMA focuses on simplifying the parts while evaluating assembly. The process works on all possible solutions to improve several aspects such as robustness, cost reduction, and the overall design. It is advisable to apply the process during the early design stages (Chang & Wysk, 1997; Kamrani & Nasr, 2010).

Several experiments were done to investigate the planning concept of generating a different set of data for different structural design problems. Digital prototypes were experiments that were done and prototyped using FDM technology and profiles shaped (wood or paper) for manufacturing the mockups. Physical prototypes were actual models that were printed in aluminum or steel.

N-AM 1st Generation was the first stage of experimenting. The goal was to test several node designs to find the most flexible one. Then, the given node consisted of a node core connected to ball shape. The number of connecting profiles defined the number of ball connections. The model was parametrically defined, and the location of the node was geometrically optimized. The proposed node design had a limited range of design possibilities. Accordingly, additive manufacturing was integrated into the planning concept.

Highrise-tessellation tested the formation of several tessellation patterns on a free-form surface that represents a building tower. It proved the possibility to tessellate a free-form surface parametrically and generate its connecting nodes. The formation of the data set of information for the structure was done. The complexity showed on the number of profiles that could connect at a node based on the tessellation type. For the highrise-tessellation round, hollow profiles were used.

In Voronoi façade, the idea was to investigate the tessellation of a free-form surface with irregular tessellation pattern (Voronoi). The idea was to generate a light façade structure. This light structure was the result of the optimization process that searched for the lightest façade construction under the defined geometrical boundary conditions to provide the least deformation possible. Rectangular profiles were applied to test the generation of the connecting nodes. The nodes were designed as a solid volume with straight sides, which connected to the straight cut connecting profiles. The sides between the profiles were rounded to accommodate the orientation of the profile. The generated structure was not uniform, thus could not generate planar filling surfaces between the profiles.

Accordingly, Paneling-Free-Form was the next project to investigate the possibility of testing surfaces for planarity. Helmuth Pottmann states that it is hopeless to optimize arbitrary meshes towards planarity. However, it is possible to generate an approximate segmentation to a reference surface (Pottmann, Schiftner, & Wallner, 2008). Different tessellation patterns were projected over different types of free-form surfaces (single and double curved), and the optimization process was run to transform as many surfaces as possible into planarity. The project proved that it is possible to optimize free-form tessellations into planar surfaces. However, in some cases, it was not possible to optimize the geometry.

In Grid Voronoi shell roof structure, the goal was to investigate two-objective optimization on the global level and topology optimization on the local level. For the two-objective global optimization, Genetic Algorithm (GA) was applied. It is a combinatorial heuristic technique that is effective and robust to solve multicriteria-discrete optimization problems (Wu & Chow, 1994). The optimization process searched for the relation between the mass of the structure and its deformation to find the best option. Under the subjected loads and the extracted reaction forces on the structure, welded and bolted connection was investigated to determine the influence of the connecting type on the topology optimization results. The generation of the data set of information for the structure was fully defined.

In Façade Technology 1, the goal was to design and engineer a unitized complex façade element. The element was parametrically defined, geometrically optimized, structurally analyzed, and the nodes were generated and topology optimized. The remodeling of the raw optimization was done to enhance the additive manufacturing of the nodes.

Physical prototypes were done to check the liability of the collected information from Digital prototypes. Nematox II was done by Holger Strauß as one of the first physical prototypes to be 3D printed in Aluminum. N-AM 10 was the first version of N-AM 2nd generation and was 3D printed in steel. For N-AM 2nd Generation, the idea was to integrate the whole process in one running script (parametric model) from global structure generation/optimization to the local level (nodes) generation/topology optimization. The programming script had many bugs that excluded further development and a suitable specialized software was used to run the topology optimization instead. N-AM_Li3 presented the first fully designed and engineered life-size mockup. All the required structural analysis was done with respect to applicable building regulations. The mockup contains N-AM_Li3 node (patent registration number: 10 2019 219 283.2) as a first fully engineered/tested

3D printed topology optimized metal façade node for standard façade systems.

The target of this dissertation is the formulation of a new design and engineering concept for complex façade/building envelopes. These envelopes are requested to be built using standard mullion-transom façade systems in the form of stick systems, unitized systems or free-form shell structure, in combination with 3D printed topology optimized metal nodes. This design and engineering concept is defined in Li3-Method.

Li3-Method consists of two main stages (digital conception, and manufacturing conception). These stages represent several phases through the planning process. The method is based on using a basic shape that is free to define by the designer. Then, the transformation of this form into an elemental parametric model. The used 3D metal printer and the metal powder/alloy helps the generation of the mechanical properties and the material model needed for the engineering of the structure. At this stage, N-AM_Li3 node can be fully engineered and tested. The rest of the nodes are designed and analyzed according to the defined simulation model. With this information, and with any complementary tests that may be required based on applied regulations, an approval in individual case can be obtained, and the logistics are defined based on the related environment of the project.

Li3-Method presents a fully functioning design and engineering method that is complemented with a patented 3D printed topology optimized metal node "N-AM_Li3".

The method proved functional only for the used systems, optimization objectives, 3D printer machines, metal powder, and selected structures that were investigated in the course of this dissertation. The method is suitable to be upgraded and further enhanced with new parameters that help improve the overall of the planning method.

7.2 Outlook

Further investigation for different 3D printing machines and alloys is needed to try to unify a material model that is not machine-related. This could be the first step to norm the material behavior of additively manufactured elements. However, this only shortens the manufacturing conception phase by reducing testing. The rules of design for additive manufacturing have to be followed.

The parametric/digital generation process of the node is clear and the logical development is the integration of these building stones in one software that is dedicated to applying the method to construct complex façade/building envelopes. The proposed software is new idea for software development companies that are interested in offering software application to the façade/building sector.

The testing of the node needs further investigation to determine all related further testing that is needed to fulfill the requirements of the approval of individual cases.

The method needs to be tested with different façade systems to determine further geometrical enhancement for the node.

More complex geometries need to be tested to ensure the suitability of the method and defines its limits.

The topology optimization of the nodes reduces the overall weight of the structure. Accordingly, the global optimization phase in Li3-Method can be iteratively run, which offers a possible improvement in the utilization of the structural elements. This improvement can impact the economic value of the structure and its efficiency. Running the global optimization phase iteratively needs further investigation.

References

Additively. (2018). Laser Melting. Retrieved from https://www.additively.com/en/learn-about/laser-melting

Alber, D., & Gabriel, K. (1983). *Zugbeanspruchte Stabtragwerke-systematische Darstellung ihrer Eigenschaften*: Werner Dusseldorf.

Amato, K., Gaytan, S., Murr, L., Martinez, E., Shindo, P., Hernandez, J., . . . Medina, F. (2012). Microstructures and mechanical behavior of Inconel 718 fabricated by selective laser melting. *Acta Materialia, 60(5),* 2229-2239.

Arciszewski, T. (1984). *Design of joints in steel space structures.* Space Structures, Third International Conference.

Baccarini, D. (1996). The concept of project complexity—a review. *International journal of project management, 14(4)*, 201-204.

Barnes, M., et al. (2000). *Widespan Roof Structures*, Thomas Telford.

Bäck, T., Fogel, D. B., & Michalewicz, Z. (1997). *Handbook of evolutionary computation:* CRC Press.

Bäck, T., & Schwefel, H.-P. (1993). An overview of evolutionary algorithms for parameter optimization. *Evolutionary computation, 1*(1), 1-23.

Behling, S., et al. (2010). *Innovative Design and Construction*, DETAIL.

Bendsøe, M., & Sigmund, O. (2003). Theory, methods and applications. *Topology optimization. Springer, Berlin.*

Bendsøe, M. P., & Sigmund, O. (2004). Topology optimization: theory, methods and applications. In: Springer.

Beyer, T., & Kuba, D. I. J. (2006). *Der wandel der fassade und ihrer funktionen im 20. jahrhundert*: Till Beyer.

Billington, D. P. (2003). The Art of Structural Design. A Swiss Legacy.

Büscher, M., Christensen, M., Hansen, K. M., Mogensen, P., & Shapiro, D. (2009). Bottom-up, top-down? Connecting software architecture design with use. In Configuring User-Designer Relations (pp. 157-191): Springer.

Chang, T.-C., & Wysk, R. A. (1997). *Computer-aided manufacturing*: Prentice Hall PTR.

Cook, W. J., Cunningham, W., Pulleyblank, W., & Schrijver, A. (2009). Combinatorial optimization. *Springer Oberwolfach Rep., 5*(4), 2875-2942.

A. Mohsen, *Design to Manufacture of Complex Building Envelopes*, Mechanik, Werkstoffe und Konstruktion im Bauwesen 56, https://doi.org/10.1007/978-3-658-30204-7

Dally, J. W., & Riley, W. F. (1965). Experimental stress analysis.

Drass, M., Schwind, G., Schneider, J., Kolling, S. J. G. S., & Engineering. (2018). Adhesive connections in glass structures—part I: experiments and analytics on thin structural silicone. 3(1), 39-54.

Ehrgott, M. (2005). *Multicriteria optimization* (Vol. 491): Springer Science & Business Media.

Facchini, L., Magalini, E., Robotti, P., Molinari, A., Höges, S., & Wissenbach, K. (2010). Ductility of a Ti-6Al-4V alloy produced by selective laser melting of prealloyed powders. *Rapid Prototyping Journal, 16(6)*, 450-459.

Fogel, D. B. (2006). *Evolutionary computation: toward a new philosophy of machine intelligence* (Vol. 1): John Wiley & Sons.

Frettlöh, D. (2002). *Nichtperiodische Pflasterungen mit ganzzahligem Inflationsfaktor.* (Dissertation). Universität Dortmund,

Gardner, M. (1997). *Penrose Tiles to Trapdoor Ciphers: And the Return of Dr Matrix:* Cambridge University Press.

Gibson, I., Rosen, D. W., & Stucker, B. (2010). *Additive manufacturing technologies : rapid prototyping to direct digital manufacturing.* Berlin: Springer.

Grefenstette, J. J. (1986). Optimization of control parameters for genetic algorithms. *IEEE Transactions on systems, man, and cybernetics, 16*(1), 122-128.

Groover, M. P. (2010). "Fundamentals of Modern Manufacturing: Materials, Processes, and Systems, John and Wiley and Sons." *Inc., Danvers, MA.*

Grunbaum, B., & Shephard, G. (1987). Tilings and Patterns, W H. In: Freeman and Company.

Haupt, R. L. (2000). *Optimum population size and mutation rate for a simple real genetic algorithm that optimizes array factors.* Paper presented at the IEEE Antennas and Propagation Society International Symposium. Transmitting Waves of Progress to the Next Millennium. 2000 Digest. Held in conjunction with: USNC/URSI National Radio Science Meeting (C.

Hauser, C., Childs, T., Dalgarno, K., & Eane, R. (1999). *Atmospheric control during direct selective laser sintering of stainless steel 314S powder.* Paper presented at the Proceedings of Solid Freeform Fabrication Symposium 1999.

Heesch, H., & Kienzle, O. (1963). *Flächenschluss: System der Formen lückenlos aneinanderschliessender Flachteile* (Vol. 6): Springer-Verlag.

Heinrich, F. (2014). Parkettierungen I. Universität Braunschweig http://www.mathematikinformation.info/pdf/MI39Heinrich.pdf *(besucht im Juni 2019).*

Helmut Pottmann, A. A., Michael Hofer, Axel Kilian. (2010). *Architekturgeometrie* Springer Verlag.

Hwang, K. J. (2010). *Advanced investigations of grid spatial structures considering various connection systems.*

ISO, D. J. B.-V., Berlin. (2017). 6892-1 2009 Metallische Werkstoffe-Prüfverfahren bei Raumtemperatur.

ISO 6892-1. (2019) *Metallic Materials* - Tensile Testing - Part 1: Method of test at room temperature. International Organization for Standardization.

Jones, R. E., & Deckro, R. F. (1993). The social psychology of project management conflict. *European journal of operational research*, 64(2), 216-228.

Jørgensen, B. (2005). Designing to target cost: one approach to design/construction integration. Designing Value, 311-319.

Kamrani, A. K., & Nasr, E. A. (2010). *Engineering design and rapid prototyping:* Springer Science & Business Media.

Karthik Alagarsamy, N. K., Atif Mohammad, Subhash Banerjee, Hai-Chao Han, Rajiv S Mishra, Aleksandra Fortier. (2016). Computational Modeling of Stent Implant Procedure and Comparison of Different Stent Materials. *J Biomed Eng, 1*, 1-10. doi:10.17303/jber.2016.101

Kaye, B. H. (2008). *A random walk through fractal dimensions*: John Wiley & Sons.

Klein, T. (2013). *Integral Facade Construction: Towards a new product architecture for curtain walls,* TU Delft.

Knaack, U., et al. (2014). *Façades: principles of construction,* Birkhäuser.

Knaack, U., Klein, T., Bilow, M., & Auer, T. (2014). Façades: principles of construction: Birkhäuser.

Kniepkamp, M., Harbig, J., Seyfert, C., & Abele, E. (2018). TOWARDS HIGH BUILD RATES: COMBINING DIFFERENT LAYER THICKNESSES WITHIN ONE PART IN SELECTIVE LASER MELTING. Retrieved from http://sffsymposium.engr.utexas.edu/sites/default/files/2018/187%20TowardsHighBuildRatesCombiningDifferentLaye.pdf

Kondic, A. (2015). *Parametrisches Skript zur Generierung und Berechnung von Stabnetzen auf Freiformflächen.* (Diplomingenieur). Teschnische Universität Graz, Graz.

Kruth, J.-P., Mercelis, P., Van Vaerenbergh, J., Froyen, L., & Rombouts, M. (2005). Binding mechanisms in selective laser sintering and selective laser melting. *Rapid Prototyping Journal, 11(1)*, 26-36.

Kühnel, W. (2013). *Differentialgeometrie: Kurven-Flächen-Mannigfaltigkeiten* (Vol. 6): Springer Fachmedien Wiesbaden.

Lan, T. (1999). Space frame structures, structure engineering handbook, CRC Press LLC, Boca Raton Google Scholar.

Lawley, A. (1978). Preparation of metal powders. *Annual Review of Materials Science, 8*(1), 49-71.

Lloyd, S. (1982). Least squares quantization in PCM. *IEEE transactions on information theory, 28*(2), 129-137.

Löber, L. (2014). Selektives Laserstrahlschmelzen von Titanaluminiden und Stahl.

Madenci, E., & Guven, I. (2015). *The finite element method and applications in engineering using ANSYS®*: Springer.

Makowski, Z. (2002). *Development of jointing systems for modular prefabricated steel space structures.* Proceedings of the international symposium. Warsaw, Poland.

Malvern, L. E. (1969). *Introduction to the Mechanics of a Continuous Medium.*

Mayine, L. Y. (2013). *Skins, Envelopes, and Enclosures: Concepts for Designing Building Exteriors*, Routledge.

McCormick, N., & Lord, J. J. M. t. (2010). Digital image correlation. *13*(12), 52-54.

Mengeringhausen, M. (1966). *Kompositionslehre räumlicher Stab-Fachwerke.* The International Conference on Space Structures. Surrey: Blackwell Scientific Publications.

Mie, G. (1908). Beiträge zur Optik trüber Medien, speziell kolloidaler Metallösungen. *Annalen der physik, 330*(3), 377-445.

Morris, P. W., & Hough, G. H. (1987). The anatomy of major projects: A study of the reality of project management.

Mulazzani, F. B.-M. (2000). Luigi Moretti. Opere e scritti. In: Milano.

Malvern, L. E. (1969). Introduction to the Mechanics of a Continuous Medium.

Murr, L., Quinones, S., Gaytan, S., Lopez, M., Rodela, A., Martinez, E., . . . Wicker, R. (2009). Microstructure and mechanical behavior of Ti–6Al–4V produced by rapid-layer manufacturing, for biomedical applications. *Journal of the mechanical behavior of biomedical materials, 2*(1), 20-32.

Nickel, A., Barnett, D., Link, G., & Prinz, F. (1999). *Residual stresses in layered manufacturing.* Paper presented at the 10th Solid Freeform Fabrication Symposium.

Norm, D. I. f. N., DIN. (1993). DIN 66131. In: Beuth.

Norm, D. I. f. N., DIN. (2016). Prüfung Metallischer Werkstoffe–Zugproben. In (Vol.

50125).

Nowatzki, T., Ferris, M., Sankaralingam, K., Estan, C., Vaish, N., & Wood, D. (2013). Optimization and mathematical modeling in computer architecture. Synthesis Lectures on Computer Architecture, 8(4), 1-144.

Oberlerchner, A. D. (2015). *Parkettierungen im Mathematikunterricht.* Retrieved from http://othes.univie.ac.at/36556/ , https://ubdata.univie.ac.at/AC12250169

Okabe, A., Boots, B., Sugihara, K., & Chiu, S. N. (2009). *Spatial tessellations: concepts and applications of Voronoi diagrams* (Vol. 501): John Wiley & Sons.

Penrose, R. (1979). Pentaplexity a class of non-periodic tilings of the plane. *The mathematical intelligencer, 2*(1), 32-37.

Peters, W., & Ranson, W. J. O. e. (1982). Digital imaging techniques in experimental stress analysis. *21*(3), 213427.

Pottmann, H., Schiftner, A., & Wallner, J. (2008). *Geometry of architectural freeform structures.* Paper presented at the Symposium on Solid and Physical Modeling.

Reinhardt, K. (1918). Über die Zerlegung der Ebene in Polygone. Retrieved from http://resolver.sub.uni-goettingen.de/purl?PPN316479497

Renckens, J. (1996). "Aluminium Facades & Architecture." *TUDelft/VMRG/VAS/FAECF.*

Richard, H. A., & Sander, M. J. V. T. V., Wiesbaden. (2009). Ermüdungsrisse.

Richter, E., Fischer, H.-M., Jenisch, R., Freymuth, H., Stohrer, M., Häupl, P., & Homann, M. (2008). *Lehrbuch der Bauphysik: Schall-Wärme-Feuchte-Licht-Brand-Klima:* Springer-Verlag.

Riebensahm, P. (2013). *Prüfung metallischer Werkstoffe* (Vol. 4): Springer-Verlag.

Roberts, P. (1984). *Commercial Atomization by the Rotating Electrode Process.* Paper presented at the Symposium on Advances in Atomizing Processes, International Powder Metallurgy Confierence, Toronto, Canada.

Rozvany, G. I. (2009). A critical review of established methods of structural topology optimization. *Structural and multidisciplinary optimization, 37*(3), 217-237.

Ruth, J., & Stahr, A. (2005). Die hohe Kunst der Knotenwerte. FASSADE *Fassaden-Architektur und-technik*(3), 38-41.

Schaal, R. (1961). *Vorhangwände: curtain walls: Typen, Konstruktionsarten, Gestaltung,* GDW Callwey.

Schatt, W., Wieters, K.-P., & Kieback, B. (2007). *Pulvermetallurgie*: Springer.

Schiftner, A., Leduc, N., Bompas, P., Baldassini, N., & Eigensatz, M. (2013). Architectural geometry from research to practice: the eiffel tower pavilions. In *Advances in Architectural Geometry 2012* (pp. 213-228): Springer.

Schlaich, J., et al. (2001). "Eine verglaste Netzschale: Dach und Skulptur-DG Bank am Pariser Platz in Berlin." *Bautechnik* 78(7): 457-463.

Schulze, D. (1994). W. Schatt. Sintervorgänge–Grundlagen. VDI-Verlag GmbH, Düsseldorf 1992. 275 S., 217 Abbildungen, 7 Tabellen, 390 Literaturangaben. ISBN 3-18-401218-2, Preis DM 78,–. *Crystal Research and Technology, 29*(3), 362-362.

Schätzke, C. (2016). Stabnetze aus Textilbeton: geometrische und konstruktive Studien zur Verwendung von Textilbetonbauteilen bei einlagigen Stabwerkschalen. Universitätsbibliothek der RWTH Aachen.

Scriba, C. J., & Schreiber, P. (2010). *5000 Jahre Geometrie: Geschichte Kulturen Menschen*: Springer-Verlag.

Shankar, S., & Mayuram, M. (2008). Effect of strain hardening in elastic–plastic transition behavior in a hemisphere in contact with a rigid flat. *International Journal of solids and structures, 45*(10), 3009-3020.

Shifeng, W., Shuai, L., Qingsong, W., Yan, C., Sheng, Z., & Yusheng, S. (2014). Effect of molten pool boundaries on the mechanical properties of selective laser melting parts. *Journal of Materials Processing Technology, 214*(11), 2660-2667.

Siebel, E., & Ludwig, N. (2013). *Die Prüfung der metallischen Werkstoffe*: Springer-Verlag.

Stahr, A. (2006). Das wohltemperierte Netz, Phd-thesis in process, Weimar, Germany.

Stahr, A. (2008). *Das wohltemperierte Netz-Zum Konstruktiven Entwurf direkt verglaster Stabnetze auf Freiformflächen.* (Dissertation). Bauhaus-Universtät Weimar,

Steinke, P. (2015). *Finite-Elemente-Methode: Rechnergestützte Einführung:* Springer-Verlag.

Stephan, S., et al. (2004). "Reticulated structures on free-form surfaces." *Stahlbau* 73(8): 562-572.

Stephan, S., Sánchez-Alvarez, J., & Knebel, K. (2004). Reticulated structures on free-form surfaces. *Stahlbau, 73*(8), 562-572.

Stephan, S., Sánchez-Alvarez, J., & Knebel, K. (2004). Stabwerke auf freiformflächen. Stahlbau, 73(8), 562-572.

Strauss, H. (2013). *AM Envelope: The Potential of Additive Manufacturing for façade constructions* (Vol. 1): TU Delft.

Suhl, L., & Mellouli, T. (2009). *Optimierungssysteme: Modelle, Verfahren, Software,*

Anwendungen: Springer-Verlag.

Tedeschi, A. (2014). *AAD_Algorithms-Aided Design*. Italy: Le Penseur.

Tedeschi, A. (2014). AAD algorithms-aided design. *Parametric strategies using grasshopper. Potenza, Italy: Edizioni Le Penseur.*

Ulrich Knaack, E. K. (2018). *Bauphysik der Fassade*

Prinzipien der Konstruktion: BIRKHÄUSER.

Upadhyaya, G. S. (1997). *Powder metallurgy technology:* Cambridge Int Science Publishing.

Van Elsen, M. (2007). Complexity of Selective Laser Melting: a new optimisation approach.

Westermann, T. (2010). *Modellbildung und Simulation: mit einer Einführung in ANSYS:* Springer-Verlag.

Wilhelm, M. (1999). Direktes selektives laser sintern einkomponentiger metallischer werkstoffe. *Dissertation, (RWTH Aachen).*

Wachsmann, K. (1959). *Wendepunkt im Bauen*, Krausskopf-Verlag.

Willems, W. M., Dinter, S., & Schild, K. (2006). Handbuch Bauphysik Teil 1. *Vieweg, Wiesbaden.*

Williams, T. M. (1999). The need for new paradigms for complex projects. *International journal of project management*, 17(5), 269-273.

Wittmann, E. C. (1987). *Elementargeometrie und Wirklichkeit: Einführung in geometrisches Denken:* Springer.

Wohlers, T. (Producer). (2016, November 20). www.wohlersassociaties.com Retrieved from http://wohlersassociates.com/blog/2016/11/metals-at-formnext/

Wu, S.-J., & Chow, P.-T. (1994). Genetic algorithms for solving mixed-discrete optimization problems. *Journal of the Franklin Institute, 331*(4), 381-401.

Xie, J., Kar, A., Rothenflue, J. A., & Latham, W. P. (1997). Temperature dependent absorptivity and cutting capability of CO2, Nd: YAG and chemical oxygen–iodine lasers. *Journal of Laser Applications, 9*(2), 77-85.

Yadroitsev, I., Yadroitsava, I., & Smurov, I. (2011). *Strategy of fabrication of complex shape parts based on the stability of single laser melted track.* Paper presented at the Laser-based Micro-and Nanopackaging and Assembly V.

Appendixes

Appendix A: EOS M 290 Datenblatt

EOS M 290: etabliertes DMLS-System zur Serienfertigung von Komponenten mit homogenen Eigenschaften bei jedem Teil, jedem Bauauftrag und mit jedem System

Mit einem Bauvolumen von 250 mm x 250 mm x 325 mm ermöglicht die EOS M 290 eine schnelle, flexible und wirtschaftliche Herstellung von Metallbauteilen direkt aus CAD-Daten.

Reproduzierbare Bauteilqualität, hohe Kosteneffizenz, umfangreiches Qualitätsmanagement

- Dank seines robusten Designs und des leistungsstarken 400-Watt-Faserlasers arbeitet das System tagein tagaus leistungsfähig.
- Die außergewöhnliche hohe Strahlqualität des Laserspots und die hervorragende Detailauflösung eignen sich ausgezeichnet für die Herstellung hochkomplexer DMLS-Komponenten, die homogene Bauteileigenschaften von Teil-zu-Teil, Auftrag-zu-Auftrag und Maschine-zu-Maschine aufweisen.
- Die breiteste Palette an validierten Materialien und Prozessen, die auf dem Markt verfügbar ist, deckt alle Kundenanforderungen ab.
- Das intuitive, offene und produktive CAM-Tool EOSPRINT ermöglicht die Optimierung von CAD-Daten für EOS-Systeme. Der EOS ParameterEditor bietet Entwicklern ein umfangreiches und offenes Tool-Set und damit mehr Freiheit für anwendungsspezifische Optimierung.
- Die umfangreiche Monitoring-Suite ermöglicht die Qualitätssicherung von allen produktions- und qualitätsrelevanten Daten in Echtzeit. EOSTATE umfasst fünf verschiedene Module: System, Laser, PowderBed, MeltPool und Exposure OT (optische Tomographie).

Technische Daten EOS M 290	
Nutzbares Bauvolumen	250 mm x 250 mm x 325 mm
Lasertyp	Yb-Faserlaser; 400 W
Präzisionsoptik	F-Theta-Linse, Hochgeschwindigkeits-Scanner
Scangeschwindigkeit	bis zu 7,0 m/s
Fokusdurchmesser	100 µm
Stromanschluss	32 A / 400 V
Leistungsaufnahme	max. 8,5 kW / durchschnittlich 2,4 kW / mit Bauplattformheizung bis zu 3,2 kW
Schutzgaszufuhr	7.000 hPa; 20 m³/h
Abmessungen (B x T x H)	
System	2.500 mm x 1.300 mm x 2.190 mm
Empfohlener Aufstellraum	min. 4.800 mm x 3.600 mm x 2.900 mm
Gewicht	ca. 1.250 kg

Software

EOSTATE Everywhere, EOSPRINT inkl. EOS ParameterEditor

Werkstoffe*

EOS Aluminium AlSi10Mg, EOS CobaltChrome MP1, EOS MaragingSteel MS1, EOS NickelAlloy HX, EOS NickelAlloy IN625, EOS NickelAlloy IN718, EOS StainlessSteel CX, EOS StainlessSteel PH1, EOS StainlessSteel 17-4PH, EOS StainlessSteel 316L, EOS Titanium Ti64, EOS Titanium Ti64ELI, EOS Titanium TiCP Grade 2

Optionales Zubehör

EOSTATE Monitoring Suite (EOSTATE Laser, EOSTATE PowderBed, EOSTATE MeltPool, EOSTATE Exposure (OT), Comfort Powder Module, IPCM-M extra, IPCM-M pro**, Nassabscheider, Strahlkabine

* Weitere Werkstoffen auf Nachfrage
** Befindet sich in der Entwicklung.

EOS GmbH
Electro Optical Systems
Hauptniederlassung
Robert-Stirling-Ring 1
D-82152 Krailling bei München
Tel.: +49 89 893 36-0
Fax: +49 89 893 36-285

EOS Niederlassungen

EOS France
Tel.: +33 437 49 76 76

EOS Greater China
Tel.: +86 21 602307 00

EOS India
Tel.: +91 44 39 64 80 00

EOS Italy
Tel.: +39 02 33 40 16 59

EOS Korea
Tel.: +82 2 63 30 58 00

EOS Nordic & Baltic
Tel.: +46 31 760 46 40

EOS of North America
Tel.: +1 248 306 01 43

EOS Singapore
Tel.: +65 6430 05 50

EOS UK
Tel.: +44 1926 67 51 10

www.eos.info • info@eos.info

Think the impossible. You can get it.

A. Mohsen, *Design to Manufacture of Complex Building Envelopes*, Mechanik, Werkstoffe und Konstruktion im Bauwesen 56,
https://doi.org/10.1007/978-3-658-30204-7

Appendix B: Zwick Z030 to Z100

Zwick / Roell

Zwick
Materialprüfung

Produktinformation

Tisch-Prüfmaschinen Z030 bis Z100 der AllroundLine

Z100 TEW mit laserXtens HP

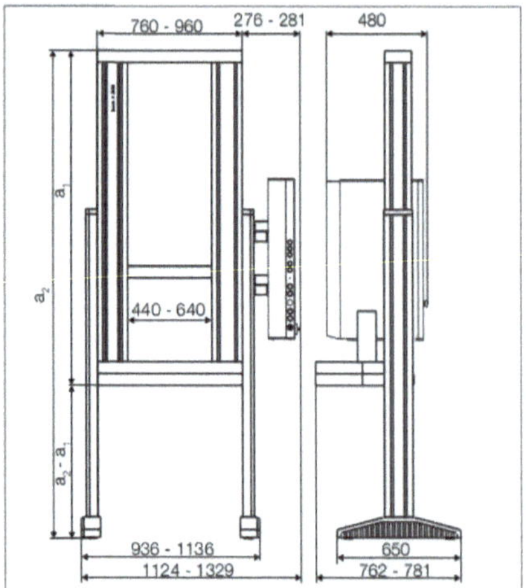

Abbildung: Z030 / Z050 TH (mit Standprofil)

Allgemeine Vorteile der AllroundLine mit der Zwick-Elektronik testControl II

Modernes Lastrahmendesign

- Der Antrieb erfolgt über wartungsfreie, digital angesteuerte AC-Antriebstechnik. In Kombination mit dem innovativen Motorfeedback-System werden auch bei kleinsten Geschwindigkeiten beste Gleichlauf-Eigenschaften erreicht.
- Die Material-Prüfmaschinen sind mit einem patentierten biegesteifen Hohlkammerprofil mit Führungszylinder ausgestattet. Die große Führungslänge und -fläche gewährleisten eine hochpräzise Traversenführung. Durch diese Kombination werden unerwünschte mechanische Einflüsse auf die Probe minimiert.
- Die im Profil integrierten T-Nuten garantieren flexible Anbaumöglichkeiten.

Hoher Bedienkomfort

- Bei der Bedienung der neuen AllroundLine Maschine steht die Ergonomie an erster Stelle.
- Die Maschine kann optimal auf ergonomische Bedürfnisse eingestellt und jederzeit modular angepasst werden.
- Eine flexible Anpassung des Arbeitsbereiches wird durch die einstellbaren Standprofile in Kombination mit der geringen Sockelhöhe erreicht. Dadurch ist auch eine rollstuhlgerechte Bedienung möglich.

Innovative Elektronik

Die neue testControl II Mess-, Steuer- und Regelelektronik bietet die ideale Voraussetzung für präzise und reproduzierbare Prüfergebnisse. Die Elektronik überzeugt durch die neue Antriebstechnologie, die hohen Messwertraten und die Modularität (Details Seite 2).

Höchste Sicherheitsansprüche

In allen AllroundLine Maschinen sind die gesetzlich vorgeschriebenen Sicherheitsanforderungen der EG Maschinenrichtlinie umgesetzt und sie erhalten eine EG-Konformitätserklärung. Es werden ausschließlich modernste Sicherheitstechniken und bewährte Industriebauteile verwendet. Ein höchstes Maß an Sicherheit für Benutzer, Prüfergebnisse, Probenmaterial und Prüfsystem sind garantiert.

Zukunftssicherheit

Dank der modularen Bauweise kann das Prüfsystem jederzeit um- und aufgerüstet werden. Des Weiteren ist die Regelelektronik testControl II kompatibel zu der zukünftigen Softwaregeneration aus dem Hause Zwick. Ersatzteile sind selbst nach Produktabkündigung noch mindestens 10 Jahre verfügbar.

PI 201 1.0916

A. Mohsen, *Design to Manufacture of Complex Building Envelopes*, Mechanik, Werkstoffe und Konstruktion im Bauwesen 56,
https://doi.org/10.1007/978-3-658-30204-7

Zwick/Roell

Zwick
Materialprüfung

Produktinformation

Tisch-Prüfmaschinen Z030 bis Z100 der AllroundLine

Daten	Wert
Lastrahmen	
Lackierung	RAL 7021 schwarzgrau und RAL 7038 achatgrau
Umgebungstemperatur	+10 ... +35 °C
Luftfeuchtigkeit (nicht betauend)	20 ... 90 %
Antrieb	
Motor	AC-Servomotor mit konzentrierten Wicklungen Hiperface® Motorfeedback-System
Motorhaltebremse	ja
Ansteuerung, Sollwertvorgabe	Digital (Echtzeit Ethernet, EtherCAT®)
Regler / Zykluszeit	adaptiv / 1000 Hz
Positionier-Wiederholgenauigkeit an der Traverse	± 2 μm
Mess- und Steuerelektronik	
Anzahl verfügbarer Steckplätze für Mess- und Steuermodule	2 synchronisierte Modulbus Steckplätze (auf 5 erweiterbar)* 1 synchronisierter PCIe Steckplatz
Kraftmessung	Klasse 0,5 / 1 je nach Kraftaufnehmer, entsprechend DIN EN ISO 7500-1, ASTM E4,
Kalkulierte Auflösung (z.B. Kraftaufnehmer in Zug- / Druckrichtung)	24 bit
Messwert-Abtastrate, intern	400 kHz
Messwert-Übertragungsrate an PC	500 Hz (optional 2000 Hz)
Nullpunktkorrektur	automatisch bei Messbeginn
Messignal-Laufzeitkorrektur für alle Kanäle	ja
Schnittstelle zum PC	Ethernet
Eco Mode	ja, automatische Abschaltung Leistungsteil (Zeit einstellbar)
CE-Konformität	ja, nach Maschinenrichtlinie 2006/42/EG
Anschlusswerte	
Netzfrequenz	50/60 Hz

* Ein hochwertiges DCSC-Messmodul für einen Kraftaufnehmer ist im Lieferumfang enthalten (belegt einen Modulbus-Steckplatz).

testControl II - Optionen, z.B.

Beschreibung	Artikelnummer
Option testControl II plus: Erweiterung der Elektronik auf 6 Steckplätze	**1008208**
2000 Hz Messwertübertragung: Erhöhung der Messwertübertragung von standardmäßig 500 Hz auf 2000 Hz. Die Messwerte werden in Echtzeit an testXpert II übertragen und verarbeitet	**057860**
Display-Fernbedienung zu testControl II für eine effektive und ergonomische Bedienung der Material-Prüfmaschine	**057984**

Optionen auf Anfrage, z.B.

- Zusatztraversen für zweiten Prüfraum
- CE-konforme, elektrisch verriegelbare Schutzeinrichtung
- Aufspannplatten
- Standprofile

Alle Daten bei Raumtemperatur.

Änderungen im Zuge der Weiterentwicklung vorbehalten.

Zwick / Roell

Zwick
Materialprüfung

Produktinformation

Tisch-Prüfmaschinen Z030 bis Z100 der AllroundLine

Typ **Artikelnummer**	Z030 TH **1004491**	Z030 TE **1004493** [1]	Z030 THW **1004492**	Z030 TEW **1004494** [1]	Z050 TH **1004495**	Z050 TE **1004498** [1]	
Lastrahmen							
Prüfkraft F_N in Zug- / Druckrichtung	30	30	30	30	50	50	kN
Prüfraumbreite	440	440	640	640	440	440	mm
Höhe des unteren Prüfraums, ohne Einbauten	1355	1755	1325	1725	1355	1755	mm
Höhe des oberen Prüfraums, ohne Einbauten (Zusatztraverse erforderlich)	1400	1800	1365	1765	1400	1800	mm
Höhe ohne Standprofile (a_1)	1714	2114	1714	2114	1714	2114	mm
Höhe mit Standprofilen (a_2)	1829 ... 2694	2229 ... 3094	1829 ... 2694	2229 ... 3094	1829 ... 2694	2229 ... 3094	mm
Breite ohne Standprofile	760	760	960	960	760	760	mm
Breite mit Standprofilen	936	936	1136	1136	936	936	mm
(und Elektronikeinheit)	(1124)	(1124)	(1324)	(1324)	(1124)	(1124)	mm
Tiefe ohne Standprofile [3][4]	618	618	618	618	618	618	mm
Tiefe mit Standprofilen [3]	730	730	730	730	730	730	mm
Gesamtgewicht mit Elektronikeinheit, ohne Standprofile	240	260	300	340	240	260	kg
Geräuschpegel bei maximaler Prüfgeschwindigkeit	64	64	64	64	61	61	dB(A)
Antrieb							
Traversengeschwindigkeit bis 110% der Prüfkraft (v_{min} ... v_{Nenn})	0,0005 ... 1000	0,0005 ... 1000	0,0005 ... 1000	0,0005 ... 1000	0,0005 ... 600	0,0005 ... 600	mm/min
Erhöhte Traversenrücklaufgeschwindigkeit (bei reduzierter Kraft)	1500	1500	1500	1500	1000	1000	mm/min
Wegauflösung des Antriebs	0,52982	0,52982	0,52982	0,52982	0,31789	0,31789	nm
Anschlusswerte							
Elektrischer Anschluss [2]	230	230	230	230	230	230	V
Leistungsaufnahme	2,3	2,3	2,3	2,3	2,3	2,3	kVA

[1] Bei dieser Material-Prüfmaschine sind die Standprofile im Lieferumfang enthalten.
[2] Elektrischer Anschluss: 230 V +/-10% (1Ph, N, PE), 50/60 Hz.
[3] Inkl. Elektronikeinheit.
[4] Inkl. Fuß (Rechteckrohr 60 x 30 mm) mit 210 mm Überstand nach vorne.

Alle Daten bei Raumtemperatur. Änderungen im Zuge der Weiterentwicklung vorbehalten.

Zwick/Roell

Zwick
Materialprüfung

Produktinformation

Tisch-Prüfmaschinen Z030 bis Z100 der AllroundLine

Typ	Z050 THW	Z050 TEW	Z100 THW	Z100 TEW	
Artikelnummer	**1004496**	**1004499** [1]	**1004500**	**1004501** [1]	
Lastrahmen					
Prüfkraft F_N in Zug- / Druckrichtung	50	50	100	100	kN
Prüfraumbreite	640	640	640	640	mm
Höhe des unteren Prüfraums, ohne Einbauten	1325	1725	1355	1755	mm
Höhe des oberen Prüfraums, ohne Einbauten (Zusatztraverse erforderlich)	1365	1765	1400	1800	mm
Höhe ohne Standprofile (a_1)	1714	2114	1769	2169	mm
Höhe mit Standprofilen (a_2)	1829 ... 2694	2229 ... 3094	1884 ... 2694	2284 ... 3094	mm
Breite ohne Standprofile	960	960	960	960	mm
Breite mit Standprofilen	1136	1136	1136	1136	mm
(und Elektronikeinheit)	(1324)	(1324)	(1329)	(1329)	mm
Tiefe ohne Standprofile [3] [4]	618	618	707	707	mm
Tiefe mit Standprofilen [3]	730	730	781	781	mm
Gesamtgewicht mit Elektronikeinheit, ohne Standprofile	300	340	450	470	kg
Geräuschpegel bei maximaler Prüfgeschwindigkeit	61	61	66	66	dB(A)
Antrieb					
Traversengeschwindigkeit bis 110% der Prüfkraft (v_{min} ... v_{Nenn})	0,0005 ... 600	0,0005 ... 600	0,0001 ... 1500	0,0001 ... 1500	mm/min
Erhöhte Traversenrücklaufgeschwindigkeit (bei reduzierter Kraft)	1000	1000	2500	2500	mm/min
Wegauflösung des Antriebs	0,31789	0,31789	0,48447	0,48447	nm
Anschlusswerte					
Elektrischer Anschluss [2]	230	230	400	400	V
Leistungsaufnahme	2,3	2,3	6	6	kVA

[1] Bei dieser Material-Prüfmaschine sind die Standprofile im Lieferumfang enthalten.
[2] Elektrischer Anschluss: 230 V +/-10% (1Ph, N, PE), 50/60 Hz; 400 V +/-10% (3Ph, N, PE), 50/60 Hz.
[3] Inkl. Elektronikeinheit.
[4] Inkl. Fuß (Rechteckrohr 60 x 30 mm) mit 210 mm Überstand nach vorne.

Appendix C: EOS StainlessSteel 316L

Material Datenblatt

EOS StainlessSteel 316L

EOS StainlessSteel 316L ist eine korrosionsresistente, auf Eisen basierende Legierung, die speziell für die Verarbeitung auf der EOSINT M280 optimiert wurde. Dieses Dokument enthält Informationen und Daten für den Bau von Teilen mit EOS SteinlessSteel 316L Pulver (EOS art.-no. 9011-0032) nach folgenden System Spezifikationen:

- EOSINT M280 400W System mit PSW3.6 und Parametersatz 316L_Surface 1.0
- EOSINT M280 200W System mit PSW3.6 und Parametersatz 316L_Surface 1.0
- EOSINT M290 400W System mit EOSPRINT 1.0 und Parametersatz 316L_Surface 1.0

Beschreibung

Bauteile aus EOS Stainless Steel 316L entsprechen in ihrer chemischen Zusammensetzung der ASTM F138 "Standard-Schmiede-Qualität für 18Cr-14Ni-2.5Mo Edelstahlteile und Draht für Chirurgische Implantate (UNS S31673)". Diese Form von Edelstahl zeichnet eine gute Korrosionsbeständigkeit aus und die Bestätigung dass es keine laugungsfähigen Substanzen in zytotoxischen Konzentrationen enthält.

Dieses Material ist ideal für den Bereich

- Lifestyle/Consumer, z.B. Uhren, Schmuck, Brillengestelle, Dekoration
- Automobil/Industrie
- Lebensmittel- und Chemieanlagen, z.B. für nicht korrodierende Bauteile
- Luft- und Raumfahrt
- Kraftwerk- und Turbinenindustrie
- Einstiegsmaterial für die Laser Sinter Technologie, z.B. Befestigungsteile, Wärmetauscher, Funktionsbauteile bei Elektronikgehäusen und Zubehör

Aus EOS Stainless Steel 316L gebaute Teile können im wie gebaut oder spannungsarm geglühten (AMS2759) Zustand maschinell sowie mit Microstrahlen weiter bearbeitet und poliert werden. Ein Lösungsglühen ist nicht notwendig, da die mechanischen Eigenschaften bereits im wie gebaut Zustand den Erwartungen entsprechen (ASTM A403). Der Einsatz von 316L Bauteilen ist in einem Temperaturbereich von 427°C bis 816°C nicht geeignet, weil hier eine Ausscheidung von Chrom-Karbiden erfolgt. Auf Grund des schichtweisen Aufbauprozesses weisen die Bauteile eine bestimmte Anisotropie auf, die sich in den mechanischen Eigenschaften zeigt.

EOS GmbH - Electro Optical Systems
Robert-Stirling-Ring 1
D-82152 Krailling / München
Telephone: +49 (0)89 / 893 36-0
Telefax: +49 (0)89 / 893 36-285
Internet: www.eos.info

A. Mohsen, *Design to Manufacture of Complex Building Envelopes*, Mechanik, Werkstoffe und Konstruktion im Bauwesen 56,
https://doi.org/10.1007/978-3-658-30204-7

Material Datenblatt

Technische Daten

Allgemeine Prozessdaten

	EOS StainlessSteel 316L
Typisch erreichbare Bauteilgenauigkeit [1], [7]	
- kleine Bauteile	ca. ±20-50 µm
- größere Bauteile	ca. ± 0.2 %
Kleinste Wandstärke [2], [7]	ca. 0,3 - 0,4 mm
Schichtdicke	20 µm
Oberflächenrauhigkeit [3], [7]	
- nach dem Bau	R_a 13 ±5 µm; R_z 80 ±20 µm
- nach dem Mikrostrahlen	R_a 5 ±2 µm; R_z 30 ±10µm
- nach dem Polieren	R_z bis zu < 1 µm (kann sehr fein poliert werden)
Volumenrate [4]	2 mm³/s (7,2 cm³/h)

[1] Erfahrungswert von Anwendern bezüglich Maßgenauigkeit typischer Geometrien, z. Bsp. ± 40 µm wenn für bestimmte Teilgruppen Parameter optimiert werden können oder ± 60 µm, wenn eine neue Geometrie zum ersten Mal gebaut wird.

[2] Mechanische Stabilität abhängig von der Geometrie (Wandhöhe usw.) und Anwendung.

[3] Auf Grund des schichtweisen Aufbaus hängt die Oberflächenstruktur stark von der Orientierung der Oberfläche ab, so zeigen beispielsweise geneigte und runde Oberflächen einen Stufen-Schritt Effekt. Die Messwerte hängen ebenfalls von der angewendeten Messmethode ab. Die hier angegeben Werte geben einen Anhaltspunkt für vertikale Oberflächen.

[4] Die Volumenrate ist ein Maß für die Baugeschwindigkeit während der Laserbelichtung, Die gesamte Baugeschwindigkeit ist abhängig von der durchschnittlichen Volumenrate, der Beschichtungsdauer (je nach Anzahl der Schichten) und anderen Faktoren wie z. Bsp. DMLS-Einstellungen.

EOS GmbH - Electro Optical Systems
Robert-Stirling-Ring 1
D-82152 Krailling / München

Material Datenblatt

Physikalische und chemische Eigenschaften der Bauteile

	EOS StainlessSteel 316L		
Materialzusammensetzung	Element	Min	Max
	Fe	Rest	
	Cr	17,00	19,00
	Ni	13,00	15,00
	Mo	2,25	3,00
	C		0030
	Mn		2,00
	Cu		0,50
	P		0,025
	S		0,010
	Si		0,75
	N		0,10
Relative Dichte bei Standardparametern	ca. 100 %		
Dichte bei Standardparametern	min. 7,9 g/cm³		

EOS GmbH - Electro Optical Systems
Robert-Stirling-Ring 1
D-82152 Krailling / München

Material Datenblatt

Mechanische Eigenschaften der Bauteile (bei Raumtemperatur) [7]

	wie gebaut
Maximale Zugfestigkeit [5]	
- in horizontaler Richtung(XY)	640 ± 50 MPa
- in vertikaler Richtung (Z)	540 ± 55 MPa
Streckgrenze, Rp0.2% [5]	
- in horizontaler Richtung (XY)	530 ± 60 MPa
- in vertikaler Richtung (Z)	470 ± 90 MPa
Young's modulus [5]	
- in horizontaler Richtung (XY)	typ. 185 GPa
- in vertikaler Richtung (Z)	typ. 180 GPa
Bruchdehnung [5]	
- in horizontaler Richtung (XY)	40 ± 15 %
- in vertikaler Richtung (Z)	50 ± 20 %
Härte [6]	typ. 89 HRB

[5] Erstellung und Test der Prüfstäbe nach ISO 6892/ASTM E8M, proportionale Prüfkörper, Durchmesser des Querschnittbereichs 5mm (0,2inch), Messlänge 4D = 20,0mm (0,79 inch), Stressrate 10MPa/s, Belastungsgeschwindigkeit im Kunststoffbereich 0.375 1/min.

[6] Rockwell Härte (HRB) Messung nach EN ISO 6508-1 auf geschliffener Oberfläche.

[7] Hinweis: diese Eigenschaften wurden auf einer EOSINT M 280-400W ermittelt. Prüfkörper von folgende Maschinentypen EOSINT M 280-200W und EOS M 290-400W erreichen vergleichbare Werte.

EOS GmbH - Electro Optical Systems
Robert-Stirling-Ring 1
D-82152 Krailling / München

Material Datenblatt

Abkürzungen

typ.	typisch
min.	minimum
ca.	circa
wt	Gewicht

Die angegebenen Werte gelten für Materialien, die mit EOSINT M 280 Systemen gemäß der aktuellen Vorgaben (inklusive der aktuellen Prozesssoftware PSW und materialabgestimmter Hardware) und Bedienungsanweisungen verarbeitet wurden. Alle Werte sind Näherungswerte. Soweit nicht anders angegeben, beziehen sich die angegebenen mechanischen und physikalischen Eigenschaften auf Standard Bauparameter und Testbauteile, die in vertikaler Orientierung gebaut wurden. Sie hängen von den Bauparametern und Baustrategien ab, die vom Anwender je nach Zielanwendung angepasst werden können.
Die Angaben entsprechen unserem Kenntnis- und Erfahrungsstand zum Zeitpunkt der Veröffentlichung. Sie bilden allein keine ausreichende Grundlage für eine Bauteilauslegung.

EOS GmbH - Electro Optical Systems
Robert-Stirling-Ring 1
D-82152 Krailling / München

Appendix D: Job quality report 180, 90

Job Quality Report

General Information

Job Date	**March 16th 2018**	Job Name	DEV_180315_Alamir_1_20180316_124306	Job Rating:
User	**eosint**	Job ID	**SI3034201803161243339**	**ERR**
Machine	**SI3034**	Machine Type	**EOS M290_400W**	
HCS Ver.	**2.6.25.0**			

Job Diagnostics

0	Events
0	User Interrupts
0	Parts modified
12/12	Parts built
n/a	Maintenance Status

Process Conditions

OK	O2 Concentration	**n/a**	Laser Power Monitoring
OK	Building Platform Temp.	**OK**	Laser Temperature
OK	Humidity Process Chamb.	**OK**	Humidity Room
!	Building Platform Pos.		

Some problems were detected. Quality may be affected. Please read through the recommendations.

Durations

Total	**15 h 21 min**
Warmup	**n/a**
Building	**15 h 21 min**
Cooldown	**n/a**

Heights

Job Start Z	**0.040 mm (0.040 mm + 0 mm)**
Job End Z	**8 mm (8 mm + 0 mm)**
Job Height	**7.960 mm (398 Layers)**
Built Height	**7.960 mm (398 Layers)**

HWI Settings

Beam offset	**-999.000 mm**	Building Platform Temp.	**80.00 °C**	Frame Temp.	**n/a**
O2 Concentration	**0.17 %**	Process Chamber Temp.	**n/a**		

Job Scaling

X-Scaling	**0.000 %**
Y-Scaling	**0.000 %**
Z-Scaling	**0.000 - 0.000 %**

Part Summary

Parts in Job	**12**	Built with Errors	**0**
Parts built	**12**	Built with Warnings	**0**
Parts not built	**0**	Built with Quality Warning	**0**

Additional Information

1	Max. deviation (20 µm) from target value of building platform position is outside tolerance range (+- 5 µm).

A. Mohsen, *Design to Manufacture of Complex Building Envelopes*, Mechanik, Werkstoffe und Konstruktion im Bauwesen 56,
https://doi.org/10.1007/978-3-658-30204-7

Appendix E: Job quality report 0 & 45

Job Quality Report

eos

General Information

Job Date	**July 6th 2018**	Job Name	DEV_180705_Alamir_20180706_122307	Job Rating:
User	**eosint**	Job ID	**SI303420180706153334**	**ERR**
Machine	**SI3034**	Machine Type	**EOS M290_400W**	
HCS Ver.	**2.6.25.0**	Job Parameter	***DEV 316L_1 0.020 mm***	

Job Diagnostics

0	Events
0	User Interrupts
0	Parts modified
42/42	Parts built
n/a	Maintenance Status

Process Conditions

OK	O2 Concentration	**n/a**	Laser Power Monitoring
OK	Building Platform Temp.	**OK**	Laser Temperature
OK	Humidity Process Chamb.	**OK**	Humidity Room
!	Building Platform Pos.		

Some problems were detected. Quality may be affected. Please read through the recommendations.

Durations

Total	**69 h 33 min**
Warmup	**n/a**
Building	**69 h 33 min**
Cooldown	**n/a**

Heights

Job Start Z	**0.040 mm (0.040 mm + 0 mm)**
Job End Z	**104 mm (104 mm + 0 mm)**
Job Height	**103.960 mm (5198 Layers)**
Built Height	**103.960 mm (5198 Layers)**

HWI Settings

Beam offset	**-999.000 mm**	Building Platform Temp.	**80.00 °C**	Frame Temp.	**n/a**
O2 Concentration	**0.17 %**	Process Chamber Temp.	**n/a**		

Job Scaling

X-Scaling	**0.000 %**
Y-Scaling	**0.000 %**
Z-Scaling	**0.000 - 0.000 %**

Part Summary

Parts in Job	**42**	Built with Errors	**0**
Parts built	**42**	Built with Warnings	**0**
Parts not built	**0**	Built with Quality Warning	**0**

Additional Information

1	Max. deviation (20 µm) from target value of building platform position is outside tolerance range (+-5 µm).

A. Mohsen, *Design to Manufacture of Complex Building Envelopes*, Mechanik, Werkstoffe und Konstruktion im Bauwesen 56,
https://doi.org/10.1007/978-3-658-30204-7

Appendix F: Trumpf 3d Printer

TRUMPF

TruPrint 3000

Additive Fertigungssysteme

Technische Daten

	TruPrint 3000
BAUVOLUMEN (ZYLINDER)	Durchmesser 300 mm x 400 mm Höhe
VERARBEITBARE WERKSTOFFE	Schweißbare Metalle in Pulverform, wie z.B.: Edelstähle, Werkzeugstähle, Aluminium-, Nickelbasis-, oder Titan-Legierungen [1]
VORHEIZUNG	200 °C
LASERQUELLE	Faserlaser 500 W
STRAHLDURCHMESSER (INDIVIDUELL EINSTELLBAR)	100 - 500 µm [2]
SCHICHTDICKE	20 - 150 µm [2]
AUFBAURATE	5 - 60 cm^3/h [3]
ANSCHLUSS UND VERBRAUCH	
ELEKTRISCHER ANSCHLUSS (SPANNUNG)	400/460 V
ELEKTRISCHER ANSCHLUSS (STROMSTÄRKE)	32 A
ELEKTRISCHER ANSCHLUSS (FREQUENZ)	50/60 Hz
SCHUTZGAS	Stickstoff, Argon
O2 KONZENTRATION	Bis zu 100 (0,01%)
BAUFORM	
ABMESSUNGEN (B X H X T)	3385 mm x 2005 mm x 1475 mm [4]
GEWICHT	4300 kg [5]

Änderungen Vorbehalten. Maßgeblich sind die Angaben in unserem Angebot und unserer Auftragsbestätigung.

A. Mohsen, *Design to Manufacture of Complex Building Envelopes*, Mechanik, Werkstoffe und Konstruktion im Bauwesen 56,
https://doi.org/10.1007/978-3-658-30204-7

Appendix G: Aluminium AlSiMg

Continental | Materialdatenblatt | Continental Engineering Services

Selektives Laserschmelzen (SLM)

Aluminium / Edelstahl / Werkzeugstahl

		AlSi10Mg		X2CrNi-Mo17-12-2 1.4404 **316L**		X3NiCoMo-Ti18-9-5 **1.2709**		
Dichte	g/cm³	2,7		7,8		8,0		
Orientierung		xy	z	xy	z	xy	z	xyz
Zugfestigkeit R_m [1]	MPa	360 ± 35	330 ± 35	630 ± 50	580 ± 35	1060 ± 100	1030 ± 100	2250 ± 100
Streckgrenze $R_{p0.2}$ [1]	MPa	190 ± 20	190 ± 20	500 ± 50	450 ± 50	700 ± 100	800 ± 50	2200 ± 100
Bruchdehnung A [1]	%	6 ± 2	4 ± 2	40 ± 15		15 ± 3		14 ± 3
Elastizitätsmodul [1]	GPa	75 ± 5	75 ± 10	150 ± 30	130 ± 30	200 ± 25	140 ± 20	180 ± 30
Härte [2]		105 ± 5 HB		22 ± 2 HRC		37 ± 5 HRC		58 ± 4 HRC
rel. Dichte [3]	%	> 99,5						
mittlere Rauheit R_a [4]	µm	12 ± 3		8 ± 2				
typische Genauigkeit [5]		± 0,1 % oder ± 50 µm						
wärmebehandelt		-	-	-	-	-	-	✓

Prüfnormen (bei Raumtemperatur): [1] (DIN EN ISO 6892-1, DIN 50125,DIN EN 10002-1), [2] DIN EN ISO 6506-1, [3] ASTM E1245, [4] (DIN EN ISO 4287-1997, DIN 4288-1998), [5] ISO 2768 (Durch Verzug kann es zu größeren Abweichungen kommen)

Dieses Dokument informiert über typische Kennwerte und Eigenschaften von additiv gefertigten Bauteilen und Strukturen. Die Angaben entsprechen unserem Kenntnis- und Erfahrungsstand zum Zeitpunkt der Veröffentlichung. Eigenschaften und Eignung hergestellter Bauteile werden hiermit weder vereinbart noch garantiert. Das Datenblatt bietet alleine keine ausreichende Grundlage für eine Bauteilauslegung. Bitte wenden Sie sich bei Fragen zu Werkstoffen und Bauteilen an adam@conti-engineering.com

http://www.conti-engineering.com
adam@conti-engineering.com

Additive Design and Manufacturing

Januar 19

A. Mohsen, *Design to Manufacture of Complex Building Envelopes*, Mechanik, Werkstoffe und Konstruktion im Bauwesen 56,
https://doi.org/10.1007/978-3-658-30204-7

Appendix H: AlSi10Mg-0403

Data sheet

RENISHAW
apply innovation™

AlSi10Mg-0403 powder for additive manufacturing

Process specification

Powder description	Aluminium alloy powder
Layer thickness	25 µm
Laser power	400 W
Additive manufacturing system	AM250

Material description

AlSi10Mg-0403 alloy comprises aluminium alloyed with silicon of mass fraction up to 10%, small quantities of magnesium and iron, along with other minor elements. The presence of silicon makes the alloy both harder and stronger than pure aluminium due to the formation of Mg_2Si precipitate.

Due to the natural formation of an oxide layer on the surface of the aluminium alloy, the material has high corrosion resistance which can be further improved by chemically anodising.

Material properties

- Low density (good for light weight components)
- High specific strength (strength to mass ratio)
- High thermal conductivity
- Very high electrical conductivity
- Responds well to post process finishing

Applications

- Automotive
- Aerospace and defence
- Electronics cooling
- Consumer goods

Generic data - wrought material

Density	2.68 g/cm³
Thermal conductivity	130 W/mK to 190 W/mK
Melting range	570 °C to 590 °C
Coefficient of thermal expansion (see note 1)	20 µm/mK to 21 µm/mK

Note 1 In the range of 20 °C to 100 °C.

Note 2 Stress relieved at 300 °C ±10 °C for 2 hr, air cooled.

Note 3 Tested at ambient temperature by Nadcap and UKAS accredited independent laboratory. Test ASTM E8. Machined before testing.

Note 4 Tested to ASTM E384-11, after polishing.

Note 5 Tested to JIS B 0601-2001 (ISO 97). As built after bead blasting.

For further information or support please contact Renishaw or visit **www.renishaw.com/additive**

A. Mohsen, *Design to Manufacture of Complex Building Envelopes*, Mechanik, Werkstoffe und Konstruktion im Bauwesen 56, https://doi.org/10.1007/978-3-658-30204-7

Renishaw plc
Stone Business Park
Brooms Road, Stone
Staffordshire, ST15 0SH
United Kingdom

T +44 (0)1785 285000
F +44 (0)1785 285001
E uk@renishaw.com
www.renishaw.com

RENISHAW
apply innovation™

Composition of powder

Element	**Mass (%)**
Aluminium	Balance
Silicon	9.00 to 11.00
Magnesium	0.25 to 0.45
Iron	< 0.25
Nitrogen	< 0.20
Oxygen	< 0.20
Titanium	< 0.15
Zinc	< 0.10
Manganese	< 0.10
Nickel	< 0.05
Copper	< 0.05
Lead	< 0.02
Tin	< 0.02

Mechanical properties of additively manufactured components

	As Built	**Stress relieved** (see note 2)
Tensile strength (UTS) (See note 3)		
Horizontal direction (XY)	442 MPa ±6 MPa	334 MPa ±1 MPa
Vertical direction (Z)	417 MPa ±27 MPa	339 MPa ±6 MPa
Yield strength (see note 3)		
Horizontal direction (XY)	264 MPa ±2 MPa	211 MPa ±2 MPa
Vertical direction (Z)	206 MPa ±6 MPa	174 MPa ±4 MPa
Elongation at break (see note 3)		
Horizontal direction (XY)	9% ±1%	9% ± 2%
Vertical direction (Z)	6% ±2%	4% ±1%
Modulus of elasticity (see note 3)		
Horizontal direction (XY)	71 GPa ±5 GPa	71 GPa ±2 GPa
Vertical direction (Z)	68 GPa ±2 GPa	66 GPa ±3 GPa
Hardness (Vickers) (see note 4)		
Horizontal direction (XY)	119 HV0.5 ±5 HV0.5	103 HV0.5 ±5 HV0.5
Vertical direction (Z)	123 HV0.5 ±2 HV0.5	98 HV0.5 ±5 HV0.5
Surface roughness (R_a) (See note 5)		
Horizontal direction (XY)	5 μm to 9 μm	
Vertical direction (Z)	7 μm to 9 μm	

For worldwide contact details, please visit www.renishaw.com/contact

 Issued: 07.2015

FSC
www.fsc.org
MIX
Papier | Fördert
gute Waldnutzung
FSC® C083411

Kolibri 360 GmbH
Ferdinand-Jühlke-Straße 7
99095 Erfurt, Deutschland
produktsicherheit@kolibri360.de